Apollo 11 – The Real Story

Recognize the patterns behind this NASA hoax and today's fake news

1st edition (1.0)   June, 2024
ISBN: 9798323055807

Andreas Märki

Copyright © 2024

Andreas Märki, Föhrenstrasse 9, CH-8703 Erlenbach ZH

https://apollophotos.ch/   a.m.maerki@bluewin.ch

All rights, including those of translation, reprinting and reproduction of the work or parts thereof, are reserved. No part of this work may be reproduced in any form (photocopy, microfilm or any other process), including for the purposes of teaching, or processed, duplicated or distributed using electronic systems, without the written permission of the author.

Despite careful proofreading, errors may creep in. The author/publisher is therefore grateful for any comments in this regard. All liability is excluded, all rights reserved.

Contents:

1 Introduction 4

2 Development of Rocket Technology and Space Travel 9

3 The Apollo Program of the USA 20

    3.1 General Information about the Apollo Program 20

    3.2 Preparatory Missions 21

    3.3 The Saturn V Rocket 27

4 Apollo 11 - THE Mission 37

    4.1 Launch into Earth Orbit and Flight Phase towards the Moon 37

    4.2 Swinging into Lunar Orbit 42

    4.3 Separation and Landing Maneuvers 45

    4.4 Moonwalks 55

        4.4.1 General Information about the Moonwalks 55
        4.4.2 Shape of the Landscape around the Landing Site 59
        4.4.3 Shadows from all Sides 68
        4.4.4 TV Camera mounted crooked 71
        4.4.5 Different Brightness of the Lunar Soil 75
        4.4.6 Aldrin's Face through the Sun Visor 76
        4.4.7 Life on the Moon?! 76
        4.4.8 Press Reaction 77
        4.4.9 Comment on Press Reactions and NASA Coverage 80

    4.5 Launch back from the Moon and Rendezvous with the Orbiter (CSM) 83

    4.6 Return Flight to Earth and Splashdown 98

5 Glorious Times 103

6 Manned Space Flight after Apollo 11 107

7 Benefits of the Apollo Moon Landings 119

    7.1 General Benefit 119

    7.2 Public or Scientific Benefit 119

        7.2.1 Overview of the Scientific Objectives 119

Introduction

	7.2.2	Moon Images	122
	7.2.3	Laser Distance Measurements to the Moon	123
	7.2.4	Summary of the Public Benefit	127
7.3		*Unofficial and Secret Use*	*128*
	7.3.1	NASA as Top Dog	128
	7.3.2	Technology Development	130
	7.3.3	Investment in Secret Projects	130
8	**Moon Landing, Conspiracy Theories and Wikipedia**		**134**
	8.1	*General Conspiracies*	*134*
	8.2	*How does the Manipulation*	*137*
	8.3	*Is the Moon Landing an Isolated Case?*	*147*
	8.4	*Has there ever been a Man on the Moon?*	*152*
	8.5	*Conclusion*	*156*
9	**About Me**		**160**
10	**Appendix**		**161**
	10.1	*Lunar Orbit*	*161*
	10.2	*Data of the Earth and the Moon*	*161*
	10.3	*Dimensions of the Lunar Landing Module (LM)*	*165*
	10.4	*Separation Maneuver*	*165*
	10.5	*Test Resolution to Fig. 54*	*165*
	10.6	*Confirmation that the Wide-Angle Lens was used for Live Video*	*166*
	10.7	*Easier Rendezvous Option for Gemini 11?*	*166*
	10.8	*Should the Moon move away from the Earth?*	*167*
	10.9	*Notes to this English Edition*	*168*
11	**References**		**169**

Apollo 11 – The Real Story

1 Introduction

Dear readers, I am delighted to look back with you on the beginning of the space age and, in particular, to relive the legendary journey to the Moon - and this time to take a close look with the necessary distance.

The USA made history with the Apollo 11 Moon landing in 1969. This Moon landing is still used today to show that nothing is impossible and that we are capable of much more than we think we are at first. Nevertheless, a flight to the Moon or even to Mars seems to be a difficult undertaking in today's time, more than 50 years after even 6 Moon flights were carried out almost routinely and in short succession. It is certainly exciting to take a closer look at how a lunar flight works, what it takes, what difficulties have to be overcome and what the lunar flights have brought us.

I myself experienced these Moon flights as a contemporary witness. I was 14 years old when Neil Armstrong was the first man to set foot on the Moon. The Moon landing is an event that almost every eyewitness remembers so clearly that they still know where they were early in the morning on July 21, 1969. In my lifetime, there have only been two such events: the Apollo 11 Moon landing and the September 11, 2001 attacks in New York.

Many people watched the live broadcast of the Apollo 11 Moon landing, even though it took place at night in Europe. I was with my family on the Col des Mosses, a pass crossing at 1400 m above sea level in French-speaking Switzerland on summer vacation in a cottage without a TV. My parents and siblings went to a restaurant for the early morning live broadcast to be there for this historic event, while I had to stay in bed that night with stomach cramps. I later saw the sequence of the exit and first Moon walk in the cinema in a Swiss film newsreel that concisely summarized world events in about 10 minutes every week.

Thanks to my enthusiasm for technology, I studied electrical engineering at the Swiss Federal Institute of Technology in Zurich and was then proud, probably also thanks to the Apollo Moon landings, when I was able to take up a position in the Swiss space industry in 1995. We were developing measuring equipment and optical instruments for satellites there, so I got a good insight into the European space program and also learned what is important in space travel. Then a colleague once told me that he was intensively studying the Apollo Moon landings and now wanted to know exactly whether everything really happened as described. I almost didn't listen to him and pushed

Introduction

him into the realm of conspiracy theorists, but somehow I stopped and got involved in a longer discussion. He had already read various critical books about the Moon landing, books in which explicitly proofs were searched, which were obviously not to be found. To this I objected that historical events are in principle hardly to prove and that a falsification of such an event would have been exposed long ago. But he added further that for example the published pictures did not show a real lunar landscape. This was a trigger for me to deal intensively with Apollo. I delved into the picture material, looked for sources and went to lectures about the Moon landing, which were numerous around 2009 on the occasion of the 40th anniversary. At a lecture at the University of Bern a professor who had worked in 1969 in America at NASA mentioned that at that time in the USA 30% of the people had thought the Moon landings were a fake. On my question whether he could explain this, he did not know an answer.

Even with the German astronaut Ulrich Walter I found poll results on his website concerning the first Moon landing. The question was: "Did the first Moon landing take place at all?" In 2002 36% said no, in 2004 it was already 44% who voted no. With these results, I do wonder why so many people question a historical event that is presented as fact by our mass media. Also Wikipedia and all "serious" newspapers do not leave any doubts about the authenticity of the Apollo Moon landings. Envy can not be the reason to believe that the Moon landing was only faked. On the contrary, many would certainly be disappointed if it turned out that Neil Armstrong never set foot on the Moon and all the positive affirmations were based on a fake. But then, what is it that makes so many people doubt?

By the way, it is also interesting in the above survey that specifically the first Moon landing is asked. Altogether there were 6 successful Moon landings in the years 1969-1972. If now Spiegel Online, from where the survey originates, asks only for the first one, doesn't it suggest that one could take it less exactly with the 5 following ones? And if this were so, if numbers 2 to 6 were faked, then why should one believe that the first one was real?

So in this book I will take a close look at the NASA documentation and see if there is false, inaccurate or at best confusing information among them that causes so many people to doubt. The reasons why so many people doubt are often sweeping statements. In the surveys of Spiegel Online also only the following two no-variants can be ticked: 1.

Apollo 11 – The Real Story

"No - the few pictures made everyone blind with enthusiasm." and 2. "No - NASA realized in time that the trip to the Moon is not worthwhile and invested the money differently." This survey does not ask for the real reasons; you can only click on one of 5 choices: two for a yes, two for a no and one "I don't care ...".

For me, the first Moon landing, that is, Apollo 11, is clearly the most important. Neil Armstrong is still a household name to many today, while the crews of the later Apollo missions are hardly known. No one really likes to celebrate the first car on a distant celestial body either.

I will focus on Apollo 11 in this book. Apollo 11 is, after all, the mission that celebrated its 50th anniversary in 2019. Besides, Apollo 11 seems to be the mission that is least doubted. I can confess to you that it was not until 2008 that I began to take doubts about Apollo 11 somewhat seriously. Before that, I had sometimes discussed this among colleagues or even watched a documentary on TV, but the conclusion was always clear: some weird guys came along with mostly even weirder arguments to question a mission that had taken place long ago. But with critical questions concerning Apollo12-17 it behaved with me somewhat differently. We had bought a color TV at home about 1971, so that I followed the last missions on TV. During the last missions the live TV pictures were transmitted in color, and in a very good quality. On Saturday evenings I also watched sports programs. Swiss television reported on soccer games in Valais in color. However, these games were first recorded with film cameras, the films were then developed, then played back in the studio and simultaneously recorded again by a TV camera. The result was a lousy quality. There I had asked myself nevertheless several times whether this would go probably with right things to and here, if from the Moon a perfect live quality was possible and from the "edge of Switzerland" only a very poor film quality, whereby not even live was sent. I have then always looked closely at the Apollo transmissions, whether everything goes to and fro with right things, but I have never been able to find a mistake. But I noticed how easy the astronauts took it for example when driving on the Moon. They commented blithely on the cornering stability of the Moon rover - there was no sign of any respect or nervousness when driving in a place where one is so strongly dependent on technology and never knows whether one will return safely. But any concerns I might have had quickly faded. The joy of the successful technical performance, which gave the USA a big bonus, prevailed. Not only had the U.S. won the race, but the U.S. also scored additional points over the Soviet Union with its lived

Introduction

transparency: the U.S. had announced all its steps in advance, and even the conversations between the mission control center and the astronauts could be heard live on the radio. My knowledge of English at that time was not sufficient to understand the conversations, but I still remember the tone of voice well. In contrast, one heard from the Soviet Union only afterwards when a test flight had been successful.

I will refer to the NASA documentation in this book. This means that everything I describe here can be directly verified by you. This is a huge advantage of the Apollo missions compared to other historical analyses: With the Apollo missions, the first-hand sources are openly available for inspection. So I never have to refer to a confidential statement of Mr. X to Ms. Y, but I can provide you with an exact source reference. For example, the documentation of the Apollo missions, when it comes to the actions at the Moon, is filed in the Apollo Lunar Surface Journal and is freely accessible via the Internet. I will describe the Apollo 11 mission for you using this journal, from launch to landing, and will of course look closely to see if the documentation is authentic or if there are contradictions. This makes my work transparent and easy to follow. I will keep it simple and make the description so that anyone can follow well. So you don't need to be an engineer, scientist or anything else to read and understand this book. Common sense is enough.

In the writing style I assume that the Moon landings took place as described. This makes the reading easier, because I can omit the words "allegedly" or "according to the NASA documentation". I use the reality form (indicative) and not the possibility form (subjunctive).

This should not distract from examining the subject comprehensively. Considering the possibilities of a partial or total falsification, I propose to distinguish the following three variants regarding the truth content:
    1. Transparent successes
    2. Embellished success
    3. Staging

The variant one, "Transparent successes", I define thereby as follows: This is the official version as described by NASA and also adopted by our media, also called doctrine: There were six successful lunar landings in 1969-1972; a seventh, Apollo 13, did not land on the Moon due to a defect. After a lunar orbit, the crew returned safely to Earth. So everything happened as reported. The documentation is authentic.

Apollo 11 – The Real Story

Variant two, "Embellished success": The documentation is faulty and contradictory. Nevertheless, at least Armstrong and Aldrin were on the Moon - or at least one successful manned Moon landing including return to Earth has taken place, at best also under exclusion of the public.

Variant three, "Staging": This variant states that Saturn 5 rockets were launched for the Apollo Moon landing missions, but that no manned space capsules flew on to the Moon. The documentaries about the actual Moon landings were pure staging. In particular, the pictures were taken in studios on Earth.

But before we look at the Apollo 11 mission in detail in the chapter 4 I would like to introduce you to the topic with an overview of the prehistory.

2 Development of Rocket Technology and Space Travel

In this chapter I show you chronologically important events from the first description of rocket technology to the beginning of the Apollo program. I made the selection in such a way that from as far as possible all subprograms (e.g. Mercury, Gemini, Venera, etc.) at least one mission is mentioned, so that you get a good impression of this phase. Also the race between the Soviet Union and the United States of America should become comprehensible in this way.

As sources I use freely available internet pages, mostly from NASA. Two often used start pages of NASA are:

NASA Missions A-Z: https://www.nasa.gov/missions

and

Planetary Missions:

https://nssdc.gsfc.nasa.gov/planetary/projects.html.

You can find a lot about this also in Wikipedia, which mostly refers to corresponding NASA pages.

As an additional source, I use some articles from the Zürichsee-Zeitung, a local Swiss newspaper in the Zurich region. My wife grew up there and recently, while cleaning out her parents' house, a small package of nicely bundled and tied up newspaper pages from the 1960s on the subject of space travel came to light.

I would like to start with the person who was the first to fundamentally deal with rocketry:

In 1903, the Russian Konstantin E. Tsiolkovsky, a teacher of mathematics and physics, publishes the first papers dealing with rockets and space flight. He explains the operation of a rocket engine with the basic rocket equation and shows the principle of multistage rockets. At that time, his works did not cause a sensation, but since his theories later proved to be correct, he is called the "father of space flight".

In 1926, the U.S. American Robert Goddard fires the first rocket filled with liquid propellant. It is primitive and weak, but it successfully demonstrates the principle with liquid propellants. In the USA, R. Goddard is therefore considered the father of modern rocket propulsion. NASA's first space flight complex, built near Washington in 1959, is named after him: the Goddard Space Flight Center.

On June 18, 1944, after a vertical launch, a rocket reaches space for the first time, i.e. an altitude of over 100 km. It is an Aggregat 4 large rocket with a liquid-fueled engine made by the German Army

Experimental Station at Peenemünde. This is the first man-made object to reach space. The technical director is Dr. Wernher von Braun. [215]

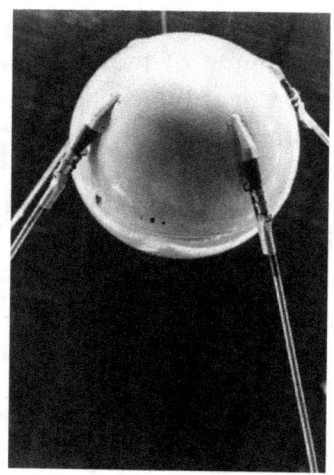

Fig. 1 Sputnik 1

On October 4, 1957, the Soviet Union launches "Sputnik" (Sputnik 1, Fig. 1) into space, the first artificial body to reach Earth orbit. Sputnik 1 is a spherical satellite with a diameter of 58 cm and a weight of 84 kg. It contains a thermometer and a radio transmitter, whose legendary beeping could be received all over the world. 92 days after the launch, in early January 1958, it burns up as it re-enters deeper layers of the atmosphere.

On October 16, 1957, twelve days later, the Swiss physicist and astronomer Fritz Zwicky launches the first object into space that would not return to Earth. Using an Aerobee rocket, he carries a metal sphere about one centimeter in diameter to an altitude of about 85 kilometers in New Mexico (USA), where a directed explosive charge accelerates the sphere to a speed of over 15 kilometers per second. This gives the bead the speed it needs to leave the Earth's gravity field forever.

On November 3, 1957, while Sputnik 1 is still in orbit, the Soviet Union launches Sputnik 2. At the same time, the first living creature is also launched into orbit, namely the dog Laika. A controlled return to Earth is neither planned nor possible. It is assumed that Laika died shortly after the launch due to excessive heat in the capsule. By the way, the word orbit means orbital path in spaceflight. Orbits around planets or around the Moon are ellipses or circles. Until you are on such an orbit, you have to accelerate. But once you are flying on this orbit with the necessary speed, you don't have to do anything and you are circling on this orbit for ever and ever. However, when you are close to the Earth, which is what all the first satellites were, you are slowed down by the upper part of the atmosphere and move slowly in a spiral towards the Earth.

Then, on January 31, 1958, the American Explorer 1 becomes the first U.S. satellite to launch into Earth orbit. It travels to 1575 miles from Earth. Dr. James Van Allen designed instruments to measure cosmic rays. The radiation belts discovered in this mission were later named

Development of Rocket Technology and Space Travel

after him.[1] The existence of these belts can be confirmed by Explorer 3 in the same year. Dear readers, please remember the discovery of these radiation belts. We will go into more detail later.

On July 29, 1958, the U.S. National Aeronautics and Space Administration (NASA) is founded. Certainly the two Sputniks accelerated this foundation.

On October 4, 1959, the Soviet Lunik 3 lunar probe is launched. For the first time, it can take photos of the side of the Moon facing away from the Earth. The Moon always shows us the same side, because its rotation is synchronous with its orbit around the Earth. The Moon's orbit is slightly elliptical and inclined; therefore, we see a bit more than half of it from Earth; in total, we can see 59% of the Moon's surface, but we never see the back 41% from Earth.

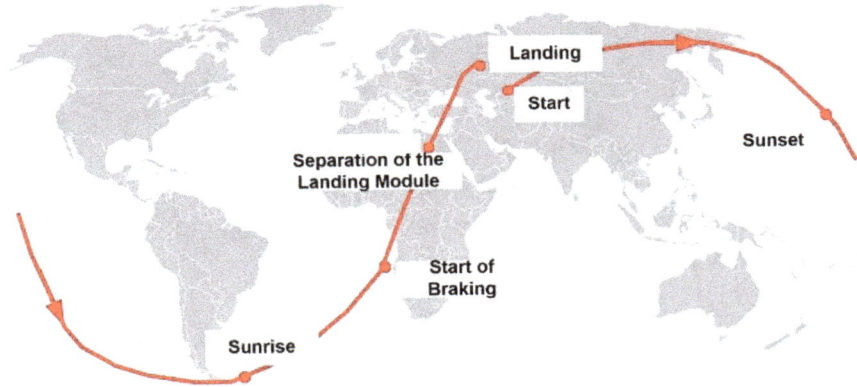

Fig. 2 Sketch of the Trajectory of Yuri Gagarin's Flight

April 12, 1961 is a bang for the buck: Yuri Gagarin becomes the first human to fly into space and orbit the Earth. The flight lasts 1 h 48 minutes. The launch is in Baikonur in Kazakhstan, where Russian rockets are still launched today, and the landing is in the Volga region. Like almost all rockets, Vostok 1 takes off towards the east in order to take advantage of the speed resulting from the Earth's rotation. The landing is then somewhat west of the launch site (Fig. 2), nevertheless Gagarin has made a complete Earth revolution, because the Earth has rotated during his flight. Interesting is the way of landing: in a height of 7,000 m the main parachute of his space capsule opens. A little later the hatch of the capsule is blown off and Gagarin is catapulted out by ejection seat. Gagarin has to separate from his seat and then his own parachute opens, with which he then lands. Even today, Yuri's Night is celebrated every April 12: these occasions are organized all over the world, either as lecture evenings or as parties. By the way, Yuri Gagarin dies on March 27, 1968, at the age of 34, during a MiG-15 training flight that ends in a crash.

On May 5, 1961, the U.S. counter with the successful flight of Mercury 3, with which Alain Shepard becomes the first American to fly into space. The counterattack is weak, however, because Mercury 3 only makes a so-called suborbital flight, i.e. it climbs to an altitude of 116.5 "statute miles" à 1.609344 km, i.e. to 187 km, and then immediately returns to Earth. The distance covered on Earth is 303 miles or 488 km. The flight path has the form of a parabola. Since it flies higher than 100 km, it has reached outer space, because this is defined from an altitude of 100 km upward. One still notices the air resistance of the atmosphere above 100 km, for example the International Space Station ISS, which flies between 300 and 400 km high, must always be raised in its flight altitude; but the pressure at 100 km altitude is less than one millionth of the pressure we have at sea level.

On February 20, 1962, the U.S. is able to catch up again: with Mercury 6, the first American, John H. Glenn Jr., flies into an Earth orbit and even circles the Earth three times. By the way, the name Mercury comes from the planet Mercury, the innermost planet or the planet with the number 1, so you can remember that there is only room for one astronaut in a Mercury capsule.

Mariner 2 (US) is launched on August 27, 1962, and flies past Venus on December 14 of the same year. For the first time, a satellite investigates a planet at close range. It discovers the high temperatures on Venus' surface of at least 425°C, higher than the melting point of lead, the composition of Venus' atmosphere of mostly carbon dioxide, and the 180° tilted axis of rotation.

Fig. 3 X-15, a Precursor of the Space Shuttle [3]

On May 15, 1963, Mercury 9, the last spacecraft in the series, is launched. The flight lasts a little more than 34 hours or 22 orbits.

On August 22, 1963, the X-15, a U.S. rocket-powered aircraft, reaches space for the second time, which by definition begins at an altitude of 100 km. Again, test pilot Joe Walker pilots the aircraft, this time to an altitude of 108 km.

Development of Rocket Technology and Space Travel

[2] The X-15 is tested by NASA test pilots from 1959 to 1968. It is used for the preparation of space flights, especially to test the behavior at high speeds. The highest speed reached is 7'274 km/h or about 2 km/s, which is 6.7 times the speed of sound. The X-15 is launched from a B-52 aircraft at an altitude of 14 km at a speed of 800 km/h, where it is then released and accelerates with its rocket engine for between 80 and 120 seconds, then flies without propulsion for typically 10 minutes and finally lands like a glider. [3] The most famous test pilot of the X-15 is Neil Armstrong. He made 7 flights from December 1960 to July 1962, reaching a maximum altitude of 63 km and a maximum speed of 6,400 km/h or 1,800 m/s, 5.7 times the speed of sound. [4]

On July 28, 1964, the U.S. Ranger 7 takes off on an unmanned flight to the Moon, where it makes a hard landing or rather an impact on July 31 as planned. During the fall, it transmits television pictures that „were 1000 times better than what could be made by Earth-based telescopes". [217]

On November 28, 1964, Mariner 4 (US) launches to Mars, where it passes on July 14, 1965, taking 22 photos. It continues toward the far side of the Sun. Braking into a Mars orbit is not yet an option at this time.

On March 18, 1965, the Soviet Voskhod 2 launches with cosmonauts Alexei Leonov and Pavel Belyayev on board. In Earth orbit, Alexei Leonov becomes the first human to disembark from his flying spacecraft. He is connected to the spacecraft only by a 4.5 m safety tether and floats in space for about 12 minutes. During re-entry he has a problem: his spacesuit has inflated and stiffened so much due to the pressure difference that he cannot get through the airlock when he tries to re-enter. Only when he emergency-releases air through a valve and thus reduces the overpressure in the suit, he can squeeze through the airlock back into the spaceship. [5]

On March 21, 1965, the last Ranger (US), number 9, launches unmanned toward the Moon. It is equipped with 6 TV cameras that send images to Earth before impact on March 24. These images are broadcast live in the USA.

On March 23, 1965, Gemini 3, the first manned flight of this U.S. series, is launched. Gemini means twins: there is room for two astronauts in these capsules. The astronauts Virgil "Gus" Grissom and John Young fly around the Earth 3 times in just under 6 hours.

Apollo 11 – The Real Story

On April 6, 1965, the U.S. places the first geostationary television satellite, Intelsat 1, affectionately known as Early Bird. Geostationary satellites orbit the Earth in a circular orbit at an altitude of 36,000 km above the equator. The orbital period is exactly one day, so these satellites are always in the same place as seen from Earth.

On June 3, 1965 at 10:15, Gemini 4 launches on a 4-day space flight. During the first orbit, an orbit maneuver is performed to reach a higher altitude. The astronauts then omit the planned subsequent rendezvous maneuver with the second rocket stage, since 42% of the fuel has already been used. So they immediately move on to the next experiment: pilot Edward White sets the pressure in his suit to 3.7 psi (3.7 psi corresponds to 0.26 bar, or about a quarter of atmospheric pressure or a pressure at 10,000 m above sea level) and the commander dumps the oxygen from the cabin; the capsule is in normal operation with pure oxygen at a pressure of 5 psi [6] (0.35 bar or air pressure at 8,000 meters above sea level) so that the astronauts can take in enough oxygen. The hatch is then opened and Edward White steps out of the Gemini capsule, still on June 3 at 14:46, becoming the first American to make a "spacewalk." He is secured with an 8 m long tether. (Fig. 4) In his right hand he holds a gas pistol for maneuvering, but it is emptied after three minutes. His visor is gold coated to protect him from the Sun's rays. The walk lasts a total of 23 minutes. [7][8]

Fig. 4 Edward White during the first US Spacewalk [9]

In the December 31 issue of the Zürichsee-Zeitung, the year 1965 is described as a record year for space travel: The Americans had now clearly taken the lead in the field of manned spaceflight, and American astronauts in particular were now in complete control of their spacecraft. The Soviet Union still had a head start in the field of super rockets - if their figures were correct.

Thus, the U.S.'s open communication has a very confidence-building effect; in contrast to the Soviet Union, which provides information only sparingly and often after the fact.

Development of Rocket Technology and Space Travel

On January 31, 1966, the Soviet Luna 9 is launched to the Moon, where it is the first satellite to land softly on a foreign celestial body. Soft landing means landing in such a way that the satellite remains intact. Luna 9 then sends photos back to Earth. This is the first soft landing ever to use a rocket engine for deceleration; small steering thrusters contribute to stabilization. Luna 9 has thus demonstrated on the Moon what the Apollo Lunar Module LM still has unspeakable trouble with on Earth, namely flying stably above the ground and also landing. We will go into more detail about this feat in the Apollo mission.

On February 22, 1966, the Soviet Cosmos 110 satellite takes off with two dogs on board. It orbits the Earth at a distance of up to 882 km and lands safely back on Earth after 22 days. The two dogs recover quickly and become darlings of the media. This mission was a biomedical experiment to study the influence of the Van Allen radiation belt on living beings.

On March 16, 1966, Gemini 8 launches with astronauts Neil Armstrong and David Scott, both on their first space flights. They perform a rendezvous or docking maneuver on an Agena rocket launched 1 hour 41 minutes earlier, designated in orbit as the target satellite. To do this, they must not only adjust the shape of its orbital ellipse, but also correct the orbital plane. After about 6 hours, they have caught up with the Agena target satellite, inspect it and dock. This is the first coupling of two spacecraft in Earth orbit. This is the first time the U.S. is now ahead in a sub-target. We will watch the rendezvous maneuver again in detail later on Apollo 11. The spacecraft with Armstrong and Scott starts to spin half an hour after docking. The astronauts are able to stop this and undock, but again their spacecraft begins to spin on its axis almost once per second. The astronauts turn off the maneuvering system and are able to bring the spacecraft under control with another system used during re-entry. On this, the landing is initiated early. [10] [11]

On May 30, 1966, Surveyor 1 (US) launches and softly lands on the Moon on June 2, 1966. It is equipped with a TV camera to take pictures of the lunar surface to assess whether the terrain is safe enough for a manned landing. This U.S. satellite also lands on the first try with the rocket engine in front.

On August 10, 1966, Lunar Orbiter 1 (US) launches toward the Moon, where, as its name implies, it orbits and photographs the Moon from August 18-29. This and subsequent Lunar Orbiter missions will map the Moon in preparation for the Apollo missions. In addition, two dosimeters are on board to measure cosmic rays on the way to the Moon. After all data are transmitted to Earth, Lunar Orbiter 1 is allowed to crash. [12] If nothing had been done, it would still be orbiting the Moon today (uncontrolled) and would be a danger for subsequent lunar missions.

On September 12, 1966, Gemini 11 takes off from Cape Canaveral for 44 orbits of the Earth. The maximum flight altitude is 1,374 km; never before was a human being so far away from the Earth. But the first orbit directly after launch is still close to Earth and slightly elliptical (160 km x 279 km). Also on this mission, a rendezvous maneuver is made to an Agena rocket launched 1 ½ hours earlier. This time it takes only 94 minutes for Gemini 11 to dock with the Agena target satellite because Gemini 11 launched exactly one orbit after the Agena rocket. The spinning away of the Earth during the one orbit, as we saw with the flight of Yuri Gagarin, Gemini can apparently compensate virtuously. As I noted with Gemini 8, we will look again at the rendezvous maneuver in more detail with Apollo 11. I would like to present another Gemini 11 feat here: still in docked condition, an astronaut goes on an outdoor mission, i.e. gets out of the Gemini 11 capsule. He detaches the loose end of a 30 m tether from the Agena target satellite and attaches it to the Gemini 11 capsule. The two spacecraft are then connected by a loose tether in addition to docking. After a few more orbital maneuvers, Gemini 11 undocks from the Agena target satellite. The two spacecraft are then connected only by the tether. Gemini 11 slowly maneuvers away from the Agena target

Fig. 5 Surveyor 3 at the Apollo 12 landing site (AS12-48-7100) [14]

satellite, stretching the tether. Gemini 11 then begins to slowly fly around the Agena target satellite. The tether remains nicely stretched, and Gemini 11 rotates around the Agena target satellite faster and faster. This swinging around of the two spacecraft causes artificial gravity on both of them, as the astronauts are constantly pushed to the outside. This is the first time artificial gravity has been demonstrated in space. After about three hours, the tether is released and the two spacecraft fly apart. [13] Artificial gravity, by the way, can be seen in science fiction movies when, for example, a large circular spacecraft is spinning so that everything is constantly pushed outward.

On April 17, 1967, Surveyor 3 launches and lands on the Moon three days later. It gained notoriety because Apollo 12 landed within walking distance and photographed it. (Fig. 5)

On April 23, 1967, the Soyuz 1 spacecraft is launched. Already at the beginning numerous difficulties occur, so that the mission is prematurely aborted after one day. During the landing the ropes of the parachute get tangled, so that the capsule including the cosmonaut V. Komarov shatters on the Earth.

On June 12, 1967, the Soviet Venera 4 spacecraft launches to Venus (Venera is Venus in Russian). On October 18, 1967, it detaches a lander that plunges into Venus' atmosphere, heat shield first. Subsequently, the fall is further slowed by parachutes, allowing the probe to measure and transmit data for 93 minutes. The data will be transmitted to Earth by two transmitter units, each transmitting one bit per second. This means that a transmitter needs at least 10 seconds to transmit, for example, only the number 1,000. At an altitude of 25 km, a temperature of 270°C and a pressure of 22 bar, the batteries then fail. The Venera 4 main probe dives into the Venusian atmosphere behind the landing module and burns up there as planned.

In the Zürichsee-Zeitung of December 29, 1967, the rapid technical progress of space research is described in the article with the headline "Harbingers of the Moon and planetary flight". In the first place the just described Venera 4 mission of the Soviet Union is mentioned as the "biggest space sensation 1967". The U.S. probe Mariner 5, which flew past Venus 36 hours after Venera 4, had been completely upstaged propagandistically by the Soviet surprise success.

In America, for the first time, the space agency was demanding larger credits for forays to Mars and Venus, but these had not been approved as a result of spending on the Vietnam War. The USA was now in danger of falling behind in the "duel of the superpowers".

On May 16, 1968, the first European research satellite ESRO-2B is launched from Vandenberg USA on a Scout launch vehicle. It studies X-rays from the Sun, cosmic rays, and the Earth's radiation belts. It burns up in the Earth's atmosphere three years later. ESRO-1A launches on October 3, 1968, with a similar mission profile. [15][16]

Apollo 11 – The Real Story

On September 15, 1968, the Soviet Zond 5 takes off for an orbit around the Moon. On board are turtles (four-toed tortoise, also called steppe tortoise or Russian tortoise) and various flies, worms and plants. On the pilot's seat sits a 175 cm tall and 70 kg heavy mannequin equipped with radiation detectors. Despite failure of two attitude control sensors, Zond 5 is able to land in the Indian Ocean and the animals are recovered alive. The turtles have lost 10% of their weight, but are active and have an appetite. We will also come back to a turtle in connection with Apollo 11.

On July 13, 1969, three days before Apollo 11, the unmanned Soviet Luna 15 launches towards the Moon. Luna 15 reaches the Moon before Apollo 11 and orbits it 52 times, i.e. during more than four days. On the afternoon of July 21, the day the Moonwalks are broadcast early in the morning in Europe, Luna 15 hits the Moon.

As a conclusion of this chapter I present a special article of the Zürichsee-Zeitung. The article was published in 1964, 5 years before the first Moon landing. The title is *The first man on the Moon* and in somewhat smaller, but still bold letters it says *Operation "Luna 1" a success - Triumph of the largest joint American-Soviet project - Lunar pioneers raise the UN flag in the "Sea of Rain" - Rejoicing all over the world*. If you look closely, right at the top of this newspaper page is the question *Will this page be correct in six years?*. A possible future Moon mission is described as an American-Soviet joint venture that had just taken place: Colonel Dimitri Spencer had been the first man to walk on the Moon, followed by Lieutenant Colonel Sergei Novikov, while Lieutenant Colonel Arthur J. Smith had continued to orbit the Moon aboard the mother ship. Spencer and Novikov then had made a series of necessary measurements on the Moon (temperature, cosmic rays, meteor activity, etc.) and then raised the flags of the United Nations, the United States, and the Soviet Union. Both the U.S. president and the Kremlin chief had sent congratulatory telegrams expressing their joy and pride at the act of the three brave space pilots. In particular, the telegram from the White House pointed out that the desire for U.S.-Soviet cooperation in the field of space exploration expressed by the assassinated President Kennedy in the fall of 1963 had found its brilliant culmination.

The stages of the flight are shown in a box. (Fig. 6).

Development of Rocket Technology and Space Travel

This report seems to me to be extremely remarkable from the following two points of view in particular:

First, there is nothing about a cold war. On the contrary, it assumes cooperation between America and the Soviet Union.

Second, cosmic rays seem to be important. On the one hand, the two space pilots carry out radiation measurements immediately after entering the Moon, and on the other hand, the box lists the successful passage through the Van Allen belt as an important stage. The author of this article obviously assumes that the reader knows that the Van Allen belt is an area of increased cosmic radiation. This belt was, after all, discovered in 1958, as described above. As I recall, "Van Allen belt" was indeed a familiar term in my youth. I mention this so clearly here because the coverage of cosmic rays and the Van Allen belt will change with the Apollo missions.

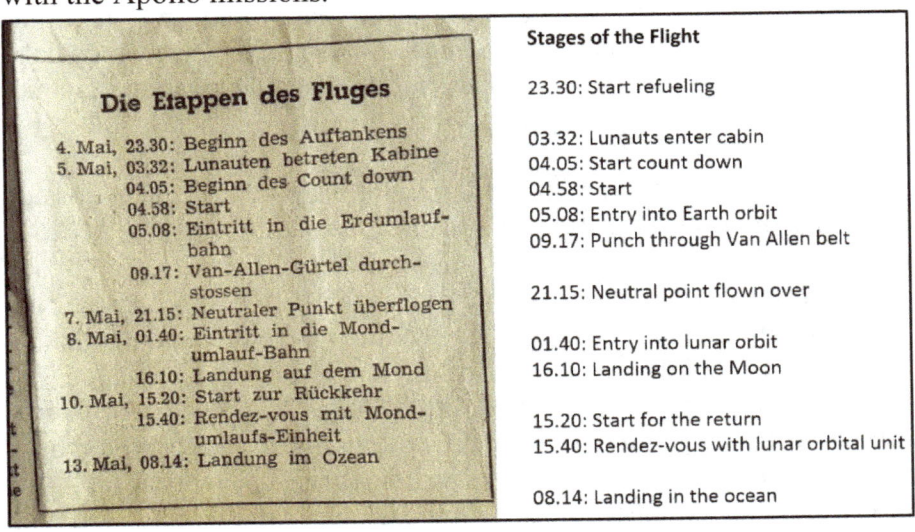

Fig. 6 Stages of the fictitious Luna 1 Flight (left: Original; right: Translation)

3 The Apollo Program of the USA

3.1 General Information about the Apollo Program

The main goal of the Apollo program is to land a man on the Moon and bring him safely back to Earth. This goal is formulated by the U.S. President John F. Kennedy on September 12, 1962 in a memorable speech at Rice University in Houston / Texas: "We choose to go the Moon ...and do the other things, not because they are easy, but because they are hard...and do all this, and do it right, and do it first before this decade is out -- then we must be bold." He had already informed the Congress about this project on May 25, 1961, and requested a budget. [17]

In 1962, the Mercury program was still running in the USA. Kennedy's speech therefore comes at a very early point in time. In the 1960s, the USA still had to make huge development steps, which John F. Kennedy also mentioned in his speech. The Mercury and Gemini programs were already preparatory missions for a Moon landing; but now this goal is publicly stated.

Even if the name "Apollo" as a mission name or part of a mission name will only appear in 1966, NASA is already thinking about a Moon landing much earlier. The mission profile, for example, is already defined in 1962: the mission should proceed in such a way that one flies with a multi-part space vehicle to the Moon, where the landing module then undocks and docks again after the landing and the relaunch.[18] Another possibility would have been to fly to the Moon with a one-piece spacecraft and land directly. Then no rendezvous maneuver would have been needed, but much more fuel would have been needed for the landing and the relaunch, because "all baggage" would have been with you all the time during the landing as well as during the relaunch.

Thus, the Apollo spacecraft has three parts: it consists of the Command Module (CM), where the astronauts will stay during the flight between Earth and the Moon; the Service Module (SM), which contains a rocket stage including propellant (when the Command Module and the Service Module are coupled together, this is called the CSM); and the Lunar Module or Landing Module (LM), which will be used to fly two of the three astronauts to the lunar surface and return to the CSM. A description of these modules is given in the chapter 3.3.

The rocket needed to carry the Apollo capsules into space and to the Moon is the Saturn rocket. There are two versions of this rocket: the smaller, two-stage Saturn IB (pronounced "one Be") for flights into Earth orbit only, and the larger, three-stage Saturn V (also known as Saturn 5) for flights to the Moon.

At the beginning, the name "Apollo" is only the name of the three-part spacecraft. Therefore the first missions with the Apollo spacecraft and the Saturn rocket are also called Apollo-Saturn-Missions (abbreviation AS). It is not until the spring of 1967 that the name "Apollo" is used as the mission name. "Apollo 1" is retroactively assigned for the planned January 1967 mission of the three astronauts who died on the ground (the description follows in Chapter 3.2), and the first Saturn V rocket launch planned for November 1967 is to be called "Apollo 4." Apollo 2 and Apollo 3 never officially existed; these names are only unofficially used sometimes. [19] So the naming is somewhat confusing, especially when you consider the beginnings of the Apollo missions. However, for the manned Apollo flights, that is, Apollo 7 onward, everything goes nicely in order. This digression on the names serves as a preparation for the next chapter.

3.2 Preparatory Missions

The Gemini program is still underway when the first unmanned Apollo Saturn mission, AS-201, lifts off from Cape Canaveral in Florida on February 26, 1966. Incidentally, the last Gemini mission is Gemini 12, which flies in November 1966.

AS-201 is an unmanned suborbital flight, i.e. a parabolic flight over a ground distance of 8,472 km with a crest altitude of 488 km. This flight tests the individual rocket stages of the Saturn IB rocket, as well as the separation of the Command Module (CM) from the Service Module (SM). These sequences work, but there are also some glitches, for example the recording of the flight data does not work during re-entry due to a short circuit. [20]

AS-203 launches on July 5, 1966 and orbits the Earth four times. AS-203 launches before AS-202; since I am proceeding chronologically, I will describe AS-203 first. This flight tests the Saturn rocket in orbit. Instead of the Apollo spacecraft, a cone-shaped attachment is mounted on the tip of the Saturn rocket.

AS-202 takes off on August 25, 1966. This flight is an unmanned suborbital flight and goes to an altitude of 1,143 km. The flight lasts 93 minutes, almost as long as Gagarin's orbit of the Earth, which took 1 hour 48 minutes. This flight mainly tests the heat shield of the Command Module. This flight goes so high to test the CM at a higher entry velocity than from a circular Earth orbit. The entry velocity of 8.9 km/s [20] is therefore greater than the circular orbital velocity of 7.8 km/s, but still less than the entry velocity for Apollo 11, which will be 11 km/s. [21]

AS-204 was to be the first manned flight of the Apollo Saturn program in February 1967. The Saturn IB rocket with the Command Module at its tip is already assembled at launch site LC-34 when a fire suddenly breaks out in the Command Module during a pre-test on January 27, 1967. The most likely cause of the fire is an electrical flashover. The Command Module has a pure oxygen atmosphere at a pressure of 16.7 psi (16.7 psi corresponds to 1.15 bar, which is 15% greater than the normal pressure at sea level). As a result of the overpressure, the inward-opening door of the Command Module cannot be opened quickly.[22]

All three astronauts involved lost their lives. Two of them were experienced astronauts: Virgil "Gus" Grissom had already been on Gemini 3 and Mercury 4, and Edward White was the first American to make a spacewalk. This (planned) mission would then be renamed Apollo 1 as described above.

NASA now classifies all tests at pure oxygen atmosphere as dangerous. However, it wants to maintain the pure oxygen atmosphere at a pressure of 5 psi (0.34 bar or pressure at 8,000 meters above sea level) in orbit. The Command Module will be modified so that air or oxygen can be pumped into the interior during tests on the ground, and the door will be modified so that it can be opened quickly during launch. Furthermore, safety measures are increased and the next planned missions are postponed and the schedule is generally stretched. However, a landing on the Moon before 1970 remains the goal. [22]

In subsequent Apollo flights, a gas mixture of 40% nitrogen and 60% oxygen is used in the Command Module during launch. In orbit, pure oxygen is used at a pressure of 5 psi (0.35 bar). [6] The fact that pure oxygen is used makes sense in that only the gas needed by the crew is carried, so that as little weight as possible is needed.

Apollo 4 (originally AS-501) is the first launch of a Saturn V rocket on November 9, 1967. Mounted on top of the Saturn 5 rocket are an unmanned Command and Service Module and a simple test model for

The Apollo Program of the USA

the LM Lunar Landing Module. Apollo 4 orbits the Earth three times and flies to an altitude of 18,000 km so that the re-entry of the Command Module into the atmosphere can then be tested under conditions that prevail on a return from the Moon, i.e. at a velocity of 11 km/s.

Apollo 5 launches on January 22, 1968; the Saturn IB rocket is used which had already been standing by during the Apollo 1 burn (AS-204). On top of the Saturn rocket sits the first Lunar Module (LM-1) as well as a cone-shaped attachment instead of the Command and Service Module. The primary objective is to operate the descent stage as well as the ascent stage of the Lunar Landing Module unmanned. All tests will take place in Earth orbit. The flight lasts a total of just under 8 hours; the mission objectives are met.

Apollo 6 launches as the last unmanned test in this series on April 4, 1968. For the second time, a Saturn V rocket (AS-502) is used. As with Apollo 4, a Command and Service Module and a simple test model for the LM Lunar Landing Module are enthroned on top of the rocket. This time, the Saturn V propulsion system test fails: two of the three Saturn V rocket stages do not function properly. But in the end, the rocket motor of the Service Module can fire for 7 minutes, catapulting the Command and Service Module to an altitude of 22,000 km. The entire flight takes just under 10 hours.

Parallel to the Apollo flights, NASA is testing various subsystems on the ground before they are to be deployed in orbit. As an example, I take the flight vehicle with which the astronauts train for the approach and landing on the Moon, the so-called LLRV (Lunar Landing Research Vehicle) (Fig. 7).

Fig. 7 Training Flight and Landing Device LLRV [23]

The astronauts make the first flight tests with the LLRV already in October 1964, before the Apollo flights. On May 6, 1968, three and a half years later or 14 months before Apollo 11, Neil Armstrong once again tries to get a grip on the LLRV. But mastering this device turns out to be more difficult than hoped. Armstrong gets into an inclined position and can only save himself with a parachute. The LLRV crashes and explodes on the ground (Fig. 8). [23]

Fig. 8 Neil Armstrong at the Parachute above the LLRV [24]

NASA calls this crash a *Rehearsal Mishap*. It would be almost 10 months before another Lunar Module (LM) could be taken into orbit (Apollo 9) for its first manned test in space.

Apollo 7 is the first manned flight of this series. On October 11, 1968, the Saturn IB rocket launches with a Command and Service Module (CSM-101), but without a Lunar Module. There are three astronauts on board. This time, the goal is to extensively test the CSM manned in orbit for nearly 11 days. After 1½ orbits, the CSM separates from the Saturn rocket, flies 15 m ahead and then makes a half turn, as will be necessary later during the flight to the Moon. By the next day, the distance between the Saturn rocket and the CSM has grown to about 80 miles. The crew locates the Saturn rocket and maneuvers the CSM back to within 20 yards of the Saturn rocket. The CSM works well, but after 15 hours of flight, one astronaut

Fig. 9 The Apollo 7 Astronauts one Hour after Splashdown [25]

gets a bad cold and also infects a second, so they have to take aspirin and decongestant medication. So the crew then survives the landing just fine, which involves an increase in pressure in the cabin, just like in an airplane. Less than an hour after landing, the crew has already arrived by helicopter at the recovery ship, where they receive congratulations on the successful 11-day flight (Fig. 9).[25]

On December 20, 1968, the Zürichsee-Zeitung runs the headline "Tomorrow Apollo 8 will be launched". The subtitle reads "On December 21, the riskiest undertaking in space travel will begin". The flight was to be the first manned trip around the Moon. Both on the way there and on the way back, several course corrections were planned, on the one hand to be able to turn precisely into a lunar orbit and on the other hand to hit the entry point into the Earth's atmosphere exactly.

NASA defined the trajectory as a *free-return trajectory*, which meant that if the engine failed on the way to the Moon, then one would make a simple loop around the Moon and could return to Earth without engine power. In America, some experts thought this project was a tour de force. But the acting director of NASA equated the risk with that taken by any test pilot testing a new type of aircraft.

As planned, Apollo 8 launches the next day. Like Apollo 7, it is manned and only equipped with a CSM (CSM-103), i.e. without a Lunar Module. But this time the launch is on a Saturn V rocket, since the trip is around the Moon. 2 hours 50 minutes after launch, the Saturn V third stage fires for 5 minutes 17 seconds, catapulting the spacecraft out of Earth orbit toward the Moon. About half an hour later, the CSM separates from the third stage rocket, which uses its remaining fuel to "clear the way" and enter solar orbit.

About 69 hours after the launch, i.e. after just under three days, Apollo 8 disappears behind the Moon and radio contact breaks off as expected. The three astronauts are the first humans to admire the far side of the Moon. 10 minutes later, the rocket engine of the CSM fires for four minutes to decelerate into a lunar orbit. This allows the crew to read Bible verses from Genesis on the occasion of a television broadcast on Christmas Eve. In total, the CSM orbits the Moon 10 times. Then it accelerates during 3 minutes 23 seconds back towards Earth. Shortly before arriving at Earth, the Command Module separates from the Service Module and the Command Module dives into the Earth's atmosphere at a speed of 11 km/s. During this process, the heat shield reaches temperatures of up to 5000°F or 2800°C.

On December 27, 1968, Apollo 8 is still on the return flight from the Moon, the Zürichsee-Zeitung appreciates the achievements of Apollo 8 and writes, now Thomas P. Stafford, the commander of Apollo 10 would have the greatest chances to be the first man on the Moon in about 20 weeks, but first Apollo 9 must try the Lunar Module manned in an Earth orbit for the first time.

The flight of Apollo 8 had brought numerous congratulations to the American space flight. On the part of Soviet scientists, the courage of the three astronauts was particularly admired. Wernher von Braun, the 56-year-old German-born American rocket scientist, had declared that he would see the establishment of permanent stations on the Moon during his lifetime.

The Soviet scientist Anatoly Blagonravov said that the American and Soviet space programs complement each other in many areas. Both sides had exchanged many results and photographs so far, and he expected the U.S. to send some results now, too.

Only in China had the mass media not taken any notice of the epoch-making Moon flight. Since 1961, it had been the custom of the Beijing leadership to pass over American and Soviet space enterprises with silence.

One day later, Apollo 8 has landed in the meantime, the Zürichsee-Zeitung writes that US President Johnson had revealed that Washington and Moscow had been in constant contact during the Apollo flight, via "hot wire", to inform the Soviets about the progress of the Apollo experiment. The newspaper further writes that there were already more than 100 people interested in flying to the Moon; no details had yet been given about the date and cost of the first Moon flight for tourists.

It was possible that the Apollo 8 team would also be selected for the first Moon landing, which was scheduled for June 1969. This had been confirmed by the director of NASA's manned space flight division.

A scientific contributor to the Zürichsee-Zeitung admires the perfection of the space venture that has just ended. As a downer, he writes at the end that the new president Nixon had announced that he wanted to withdraw huge financial resources from NASA and give them to military space flight projects. In this way, the great possibilities of technology could be rashly sacrificed to yesterday's thinking about power.

Apollo 9 launches on March 3, 1969 with a CSM and for the second time after Apollo 5 again with a Lunar Landing Module (LM-3), which was named "Spider". Although the trip is only to Earth orbit, a Saturn V rocket is used. The LM will be flown manned for the first time, and it will perform separation and docking maneuvers.

CSM and LM perform the complete separation sequence including docking maneuver as it will be performed later during the flight to the Moon. The exact description of this sequence follows in Apollo 11. An astronaut then checks the exit of the LM and climbs onto the "porch" of the LM, where he stays for more than half an hour, i.e. outside the LM. The astronauts then use the LM to practice maneuvers like those later done on the Moon: two astronauts fly the LM 113 miles away from the

The Apollo Program of the USA

CSM, blast off the LM's descent stage, and return to the CSM on the ascent stage. On March 13, 1969, after 10 days, the three astronauts land about 340 miles north of Puerto Rico in the Atlantic Ocean, and the LM's descent stage burns up in the atmosphere on March 22, 1969.

Apollo 10 takes off as a main rehearsal on May 18, 1969, heading for the Moon. When Apollo 10 is 3,570 miles from Earth after a three-hour flight, the first live color TV transmission to Earth begins: the inside of the CSM and the docking are shown. During the flight, everything happens as during the following Moon landing, except that the LM (LM-4, "Snoopy") does not touch down on the Moon. That is, the CSM orbits the Moon while the LM approaches the Moon on an elliptical orbit to within 9.7 miles or 16 km. Instead of initiating the landing as Apollo 11 later will do, the LM blasts off the descent stage, which remains in lunar orbit and probably still orbits the Moon today. Subsequently, the LM uses the ascent stage to correct its orbit so that it can rendezvous with the CSM again and dock there. In total, the mission lasts 8 days.

Fig. 10 Apollo 10 Emblem

Interesting is the emblem of Apollo 10 (Fig. 10): the LM is on its way back to the CSM after the approach to the Moon, and the engine is firing, which can be seen at the fire ray. What sounds so inconspicuous and logical here, namely that one can see from the fire stream that an engine is firing, is later described differently by NASA for the return launches from the Moon.

3.3 The Saturn V Rocket

The Saturn V (also Saturn 5) rocket is the rocket that was built to carry humans to the Moon. It was used on all Apollo missions that went to the Moon. For preparatory missions in Earth orbit, its smaller sister, the Saturn IB, was also sufficient. Nevertheless, the Saturn V was also used for test purposes on flights "only" to Earth orbit.

The designation of the Apollo 11 Saturn V rocket is AS-506, which, including the Apollo capsule and escape rocket, is 111 m high, equivalent to the height of a 36-story building, and has a launch weight of 2,800 t (tons). The thrust at launch is 3,500 t. [26] The thrust at launch must be greater than the launch weight for the rocket to lift off. If it were exactly the same, then the thrust could only keep the rocket in hover and just compensate for the Earth's gravity. For Saturn V, the thrust is 25% greater than the launch weight, which means the rocket accelerates at a quarter of the acceleration due to gravity, or ¼ g, at launch. The acceleration then increases over time as the rocket becomes lighter and lighter due to the burning of the propellant. But at launch, the acceleration is very low as described. If you watch videos of such launches, you see that the rocket is slow to get off the launch pad.

Fig. 11 Size Comparison of Saturn V, Saturn IB, Space Shuttle, Soyuz FG, Ariane 5 and Falcon Heavy

The Saturn V is the largest rocket ever built. Fig. 11 shows a size comparison of the Saturn rockets with other known rockets: the Space Shuttle flew 153 times into low Earth orbits, i.e. up to 600 km orbit altitude, between 1981 and 2011. The Soyuz FG is used for manned feeder flights to the ISS. It still takes off from Baikonur/Kazakhstan. Ariane 5 is the current European rocket that lifts off from Kouru/French Guiana, and Falcon Heavy carried a Tesla car toward Mars on its first

The Apollo Program of the USA

launch on Feb. 6, 2018. Falcon Heavy is currently (2018) the most powerful rocket.

The so-called payload of the Saturn V, i.e. what it carries into space, is 120 t for Earth orbits; to the Moon it is still 43 t. This is a record amount, but compared to the launch weight of 2,800 t, the payload for an Earth orbit is just 4% and for a lunar flight even only 1.5%. The predominant parts of a rocket are fuel tanks.

Like all large rockets, the Saturn V rocket is multi-stage: it consists of three stages. The visible nozzles at the bottom of the rocket belong to the first or lowest stage (S-1C, Fig. 12). Once all the fuel in the lowest stage is used, it is jettisoned. This makes the rocket lighter and the next stage, which is then used, has to accelerate much less mass. The second stage then uses new engines, but these are much lighter than the hull of the just jettisoned first stage, so it is worth building rockets with multiple stages. This can be compared to a cyclist who prefers to carry three small drinking bottles rather than one large one. As soon as one is empty, he throws it to the side of the road and thus has to carry less empty shell.

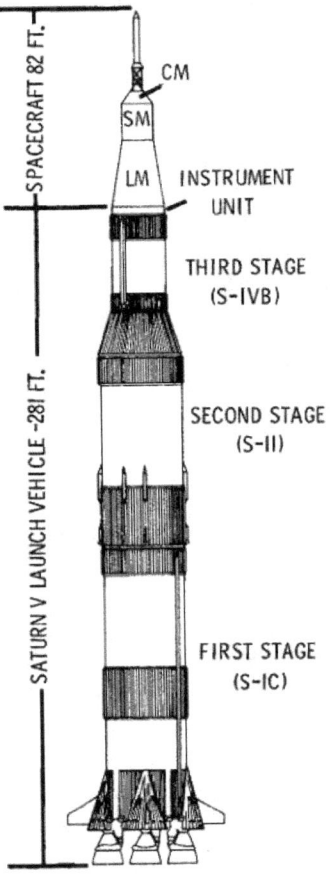

All stages are operated with liquid fuel: either liquid hydrogen or a kerosene-like liquid. The liquid fuel is injected as a gas from its tank into the combustion chamber. For it to burn, it needs oxygen. This comes from a second, separate tank and is also injected into the combustion chamber. Combustion starts spontaneously - comparable to a diesel engine, which also does not need spark plugs. Fuel and oxygen burn in the combustion chamber, escape backwards at high speed and thus produce the thrust of the rocket. Unlike an airplane, which burns its fuel kerosene with oxygen from the air, the rocket carries the necessary oxygen in a separate tank. This allows it to burn the fuel much faster than an airplane. The oxygen is carried either as

Fig. 12 Saturn V Rocket [27]

liquid oxygen, which must be cryogenically refrigerated, or as an oxygen-rich compound that does not need to be refrigerated and thus has a longer shelf life. The generic term for the oxygen or oxygen-rich compound is oxidizer.

Rockets work the same in air and in vacuum. Gases, which are hot and therefore glow, are ejected backward, and as a reaction, the rocket is accelerated forward. This is called the recoil principle. You can try this principle yourself by pushing a heavy stone forward from a standing position; because of the recoil, you then have to be careful not to fall over backwards. Mostly spaceships are "in free fall"; the rocket motor does not burn then and the spaceship flies like a stone.

The first stage (S-1C) of the Saturn V rocket has five nozzles at the bottom, each belonging to a separate (rocket) motor or engine. They are arranged the same as the five points of a die: four on the outside and one in the middle. These 5 engines burn for 2½ minutes at launch and together burn 15 tons of liquid oxygen and a special kerosene for rockets (RP-1) per second. At least one nozzle must be swiveling so that the direction of thrust can be controlled. When flying straight, for example, the thrust must be precisely aligned with the rocket's center of gravity; if it is not, it flies a curve. Of the five nozzles, the outer four swivels, while the inner one is fixed to the rocket. [28] The first stage has a diameter of 10 m and a height of 42 m. [29]

The second stage (S-II) also has 5 motors. They are arranged the same as in the first stage and burn for about 6 minutes. Liquid hydrogen and liquid oxygen are used as fuel. This stage is also 10 m in diameter; it is shorter than the first stage, still 25 m long. [27] Unlike the first stage, it no longer has stabilizing wings; nor does it need them, since the air is very thin at the altitude where it is used.

The third stage (S-IVB or S-4B) has only one engine and is narrower. Its diameter is still 6.6 m and it is 18 m long. Like the second stage, it is operated with liquid hydrogen and liquid oxygen. [30] Unlike the first two stages, the fuel supply can be interrupted and restarted. We will see where this is needed in the next chapter. This stage is also used on its smaller sister, the Saturn IB. It allows the Lunar Module or the whole Apollo spacecraft to be attached to the front; and it sits there directly on the first stage.

The Apollo Program of the USA

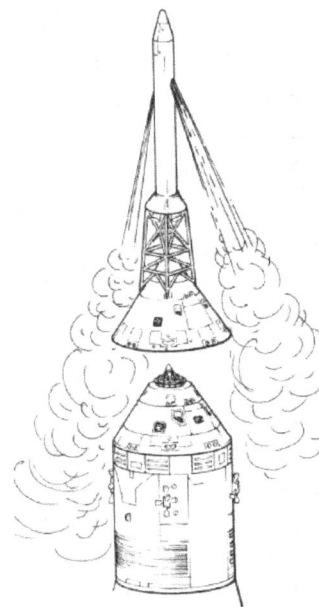

Fig. 13 Detachment of the Rescue Rocket (top) from the CSM (bottom) [31]

Above the third stage is the instrument unit (IU) of the Saturn V rocket. It is virtually the brain of the rocket and steers it through the atmosphere into Earth orbit. For this purpose, it has an inertial platform, acceleration sensors and a radio link to the ground. Its final task is to perform the acceleration maneuver toward the Moon (Translunar Injection TLI). [28]

The Saturn V payload is the three-part Apollo spacecraft consisting of the space capsule or Command Module (CM), the Service Module (SM) and the Lunar Module (LM).

At the very top there is a 10 m high rescue rocket, with which the Command Module (without Service Module) can be jettisoned together with the astronauts. This is called the Apollo Launch Escape System (LES). In the event of an emergency during launch preparations or shortly after launch, the crew could escape to safety by manually triggering the escape system. The escape system could also be triggered automatically, for example, if two of the five rocket engines failed during the first stage burn or if the Saturn V rocket became misaligned. However, the escape rocket was never needed; it was always jettisoned in a controlled manner as soon as the second stage ignited. Fig. 13 shows the rescue rocket during the normal jettison phase. The two upper rocket motors are firing.

In an emergency, the lower rocket motors would have fired (Fig. 14) - they can also be seen in Fig. 13 at the upper end of the latticework - and thus brought the Command Module to a safe distance from the Saturn V rocket and an altitude of at least 7'500 m. There, the escape rocket would have jettisoned and the Command Module could have landed normally on its parachutes. [31]

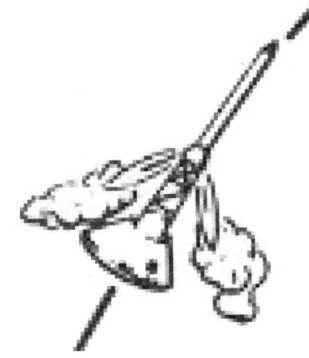

Fig. 14 Rescue Rocket in Emergency Operation with Command Module CM [31]

Apollo 11 – The Real Story

Back to the payload, i.e. the three-part Apollo spacecraft. Directly below the rescue rocket sits the Command Module (CM) with the three astronauts, which could be jettisoned together with the rescue rocket in an emergency as just described. It is connected to the Service Module (SM). The two modules remain together during almost the entire mission, forming the Command and Service Module CSM. Only shortly before landing on Earth is the Service Module disconnected.

The CSM is given the name Columbia on Apollo 11 and has the designation CSM-107.

The three astronauts sit or lie in the Command Module in such a way that they are pressed into the seat during all maneuvers in which high acceleration occurs; i.e., about the same as during the takeoff of an airplane. In this way, the blood is neither pressed into the head nor is there a blood void. The Command Module is comparable to the cockpit of an airplane: here are all controls and displays (Fig. 15). The three astronauts stay most of the time in the Command Module. Only after entering the lunar orbit two astronauts go into the lander. Until the re-ascent, the Command Module is occupied by only one astronaut, the pilot.

The Command Module weighs 5.5 t, has a diameter of 3.9 m and is 3.65 m long.

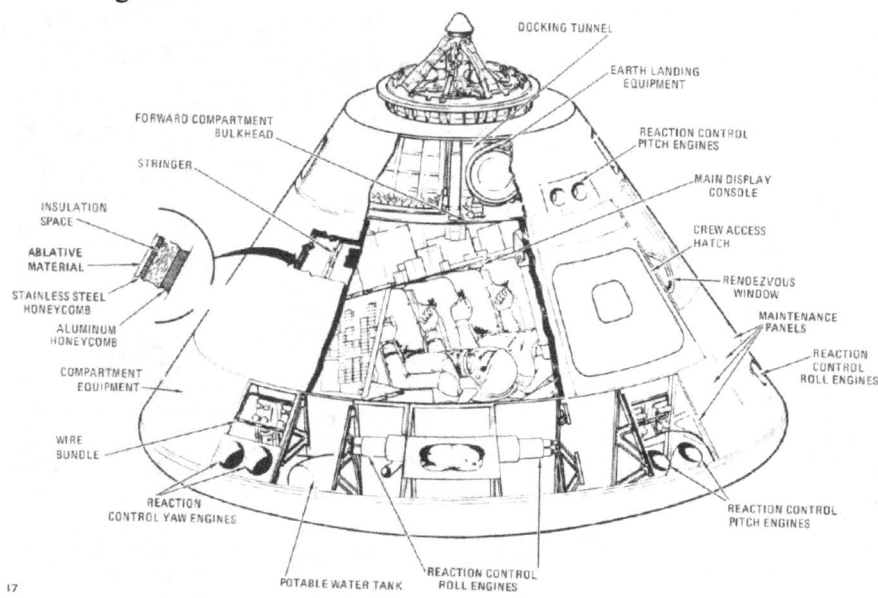

Fig. 15 Command Module CM [32]

The Apollo Program of the USA

As can be seen on Fig. 15 the Command Module has its own attitude control system: hot gas can be ejected from various openings. This works in the same way as a rocket motor with liquid propellant (here with monomethyl hydrazine) and oxidizer (dinitrogen tetroxide, N_2O_4). [32] The two components ignite as soon as they come into contact with each other. [33] The attitude control system is needed during re-entry into the Earth's atmosphere: before re-entry, the Command Module is rotated so that the aft heat shield (the larger one) is in front, and during flight through the atmosphere, the attitude control system ensures that the aft heat shield also remains in front, i.e. that no uncontrolled rotations occur.

Fig. 16 Command and Service Module CSM [36]

The Service Module SM is connected to the command module as described above. It could also be called the fourth rocket stage: its main components are the rocket motor (Service Propulsion System SPS) and the propellant tanks. The Service Module is not accessible to the astronauts; it is fully packed. There is also no passage to the bottom, because there is the protective shield. In addition to rocket propulsion, the Service Module provides other functions, for example, the antennas for communication with Earth are attached to the Service Module and it houses fuel cells that provide the electrical supply (28 volts) with the help of oxygen and hydrogen, as well as the common attitude control system: every 90° on the circumference, a total of four groups of four control nozzles (Fig. 16 and Fig. 17). The fuel and oxidizer of the attitude control system are identical to those of the Command Module. [33]

The Service Module has the same diameter as the Command Module, namely 3.9 m. It is 7.6 m long. [34]

The fuel used is aerozine 50 (hydrazine mixture, similar to kerosene) and the oxidizer is dinitrogen tetroxide (N_2O_4).[35] Liquid oxygen is therefore no longer used here as in the first three stages. Liquid oxygen enters the tanks in a deep-frozen state and can therefore only be stored for a short time. For several days to the Moon and back, a fuel is needed that can also be kept at normal temperatures.

The Lunar Module (LM) is designed for two astronauts. With it, they can first fly from a lunar orbit to a lower orbit and then land on the Moon. The rocket motor for deceleration and landing is in the lower part of the LM, called the descent stage. This remains on the Moon; for the return launch from the lunar surface, the two astronauts need only the upper part, the ascent stage, where their cockpit is also located. At Fig. 18 you can see into the narrow cockpit (crew compartment). It has no seats or similar, the two astronauts can only stand there.

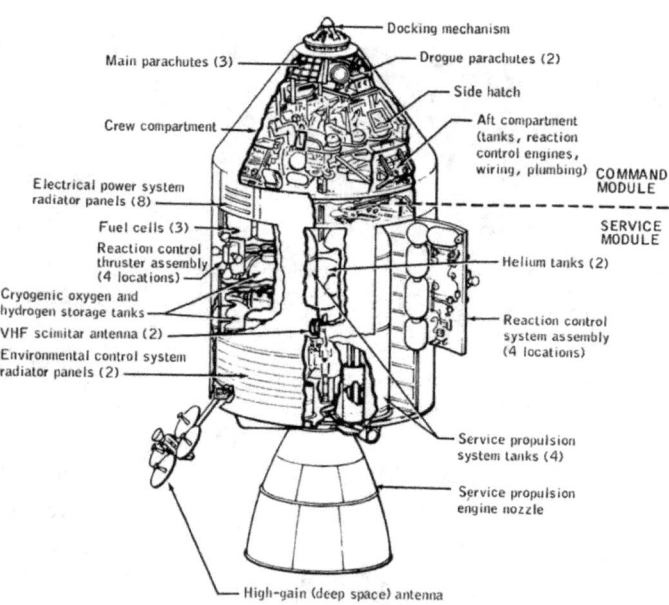

Fig. 17 Command and Service Module CSM [34]

The descent stage could also be called the first rocket stage of the Lunar Module. Its task is to decelerate during the landing phase, and when this is done, it is no longer needed and left behind. The spacecraft for the ascent thus becomes lighter and accordingly manages with less fuel.

In the Lunar Landing Module, the same propellant mixture is used for both the rocket motors and the attitude control system as for the motor of the CSM. [37] [28] Like the CSM, the Lunar Landing Module has four nozzle groups of four nozzles each, mainly to generate rotary motion and then bring it to a standstill again.

The Lunar Module LM weighs 15 t fully fueled; 8.2 t of it are propellant in the descent stage. That means, the LM as we see it on the lunar images still has a mass of about 7 t, because the propellant of the descent stage is used up. The height with extended props is 7 m; this is the height as it is shown on the lunar images. [37][38]

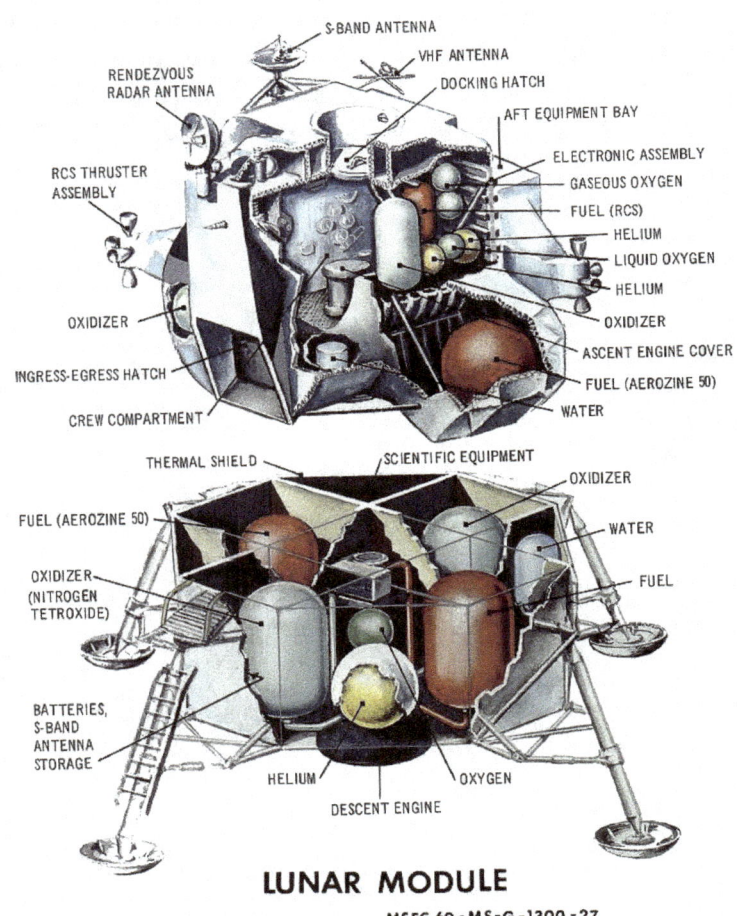

Fig. 18 Lunar Module with Ascent Stage (top) and Descent Stage (bottom)

On Apollo 11, the Lunar Landing Module (LM-5) was named Eagle. This name is immortalized in Neil Armstrong's famous phrase "The Eagle has landed".

The motor of the ascent stage is fixed, so it is not gimbaled, while the motor of the descent stage can be deflected 6° in all directions. [39][40] This means that for the ascent stage only the (weak) control thrusters can correct the orientation, which was sufficient for this rather small spacecraft with a mass between about 5 t at launch and 2.5 t when empty.

In Fig. 18 you can see several helium tanks inside the Lunar Landing Module. The helium is needed to operate the rocket engines and attitude control system: it is fed into the tanks, forcing propellant and oxidizer out of their tanks and into the combustion chambers.

4 Apollo 11 - THE Mission

4.1 Launch into Earth Orbit and Flight Phase towards the Moon

On Wednesday, July 16, 1969 in the morning at 9:32 local time, the Saturn V rocket lifts off from launch pad 39A in Cape Canaveral / Florida. The three astronauts Neil Armstrong, Edwin "Buzz" Aldrin and Michael Collins are sitting, or rather lying, in the Command Module. In 1969, there was no daylight-saving time in Europe, while the local time is given as EDT "Eastern Daylight Time". The time difference to Central Europe was therefore only 5 hours.

The first stage burns for 2 ½ minutes and takes the Saturn V rocket to an altitude of 38 miles or 60 km at a velocity of about 3 km/s. It is jettisoned there and burns up in the Earth's atmosphere. [29] [41]

The second stage then burns for about 6 minutes, propelling the rocket to an altitude of 115 miles or 185 km and a distance of 1,500 km. Now the Saturn V rocket has reached space, which by definition starts at 100 km altitude. Its speed is now already 7 km/s. [27] Thus only little is missing for the necessary speed of 7.8 km/s in the targeted circular orbit.

The necessary acceleration to put the Apollo spacecraft into the circular park orbit is provided by the third stage. This must be fired for 2½ minutes. [42] But the third stage is not yet burned out; it will be needed again shortly. Until that time comes, the astronauts can test the proper functioning of the Apollo spacecraft in the circular orbit of 191 km altitude. The astronauts are now in weightlessness. The entire launch phase to reach weightlessness in park orbit took less than 12 minutes.[21]

After almost two orbits of the Earth and a total flight time of 2 ¾ hours, the Apollo spacecraft is now to leave Earth orbit and continue towards the Moon. For this purpose, the third stage S-IVB will be ignited for the second time, this time for almost 6 minutes. [21] The necessary speed increase is greater than for the first ignition: the speed must be increased from 7.8 km/s to 10.8 km/s for the journey to the Moon. [42] The completion of this maneuver is called TLI (Translunar Injection). Incidentally, the speed of 10.8 km/s is just slightly less than the escape velocity, which is 11.0 km/s from this altitude. With 11.0 km/s a spacecraft would move away from the Earth forever; with 10.8 km/s, however, it flies on an Earth orbit, which in the case of Apollo 11 is then additionally influenced by the Moon.

Apollo 11 – The Real Story

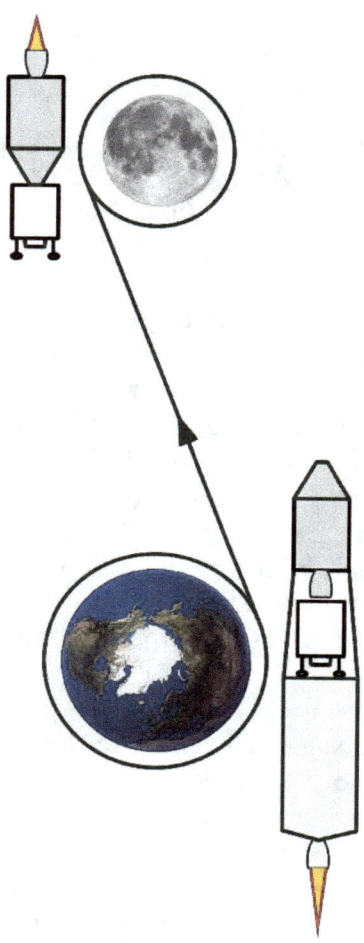

Fig. 19 Trajectory and Configuration when leaving the Earth's Orbit and Arriving at the Moon

So now the Apollo spacecraft is traveling at almost 11 km/s towards the Moon, and once again the astronauts are in weightlessness. The speed decreases continuously the further one moves away from the Earth. Only at the equilibrium point between the Earth and the Moon, that is, where the Moon's gravitational pull is equal to the Earth's gravitational pull, will the speed increase again due to the Moon's gravitational pull. By then, one will have traveled 90% of the way to the Moon.

The third stage is now burned out and has had its day. But before it can be jettisoned, the three-part Apollo spacecraft must be regrouped. In Fig. 19 you can see the simplified trajectory from the Earth to the Moon, and at the bottom right is shown how the third stage fires and accelerates the Apollo spacecraft towards the Moon. In the upper left, the braking into lunar orbit is drawn. The arrangement of the Command and Service Module CSM relative to the Lunar Landing Module is different. The CSM must be turned on the way to the Moon, so that on the one hand its rocket motor does not fire into the Lunar Module and on the other hand the Command Module with its tip comes to rest on the Lunar Module, so that the astronauts can use the passage to the Lunar Module later. It would not have been possible to arrange it this way from the beginning, because during the launch one wanted to have the possibility that the astronauts in the Command Module could have been jettisoned in case of emergency. So the Command Module had to be on top. In this constellation, as described earlier, the three astronauts are pressed into their seats during all large accelerations.

Apollo 11 - THE Mission

The turning maneuver of the Command Module is shown in Fig. 20. First, the CSM and the casing, which served to protect the Lunar Landing Module, are jettisoned from the original configuration at the very top. Then the CSM flies forward about 20 m, makes a half turn, flies back and docks with the Lunar Landing Module. For this maneuver, the small control nozzles of the attitude control system are used to control rotational, longitudinal and lateral movements.

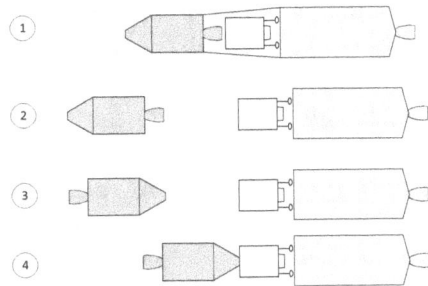

Fig. 20 Turning Maneuver of the Command and Service Module CSM

This turning maneuver is performed about half an hour after leaving the Earth's orbit at a distance from the Earth between 7'600 km and 9'800 km and takes 7 minutes. [21] One is still quite close to the Earth in terms of the whole trip; the distance from the Earth is about ¾ of a diameter of the Earth, which is 12'740 km.

About an hour later, the Apollo spacecraft and the third stage separate, and the third stage then fires briefly, as it did for Apollo 8, to enter a solar orbit so that it will certainly not get in the way of the Apollo spacecraft. [43]

Dear readers, it is now time to pause and take a closer look at the turning maneuver or, more importantly, the place where it takes place.

For this, I have to elaborate a bit. I have described in the chapter 2 that Explorer 1 discovered the Van Allen radiation belt in 1958. This belt surrounds the Earth like an inflated floating ring, and inside it there is intense "radioactive radiation" (white areas in Fig. 21). I write this in quotation marks because the radiation does not come from the decay of radioactive elements and cannot itself be radioactive. However, the radiation is the same that comes from radioactive elements. In the Van Allen radiation belt protons and electrons fly around at high speed, i.e. positively and negatively charged particles of atoms. These elementary particles are trapped there in the Earth's magnetic field. Due to their high speed they produce gamma radiation. Gamma radiation is even shorter wavelength than X-rays and correspondingly more dangerous to health.

Apollo 11 – The Real Story

On the subject of gamma radiation, I recently read an article in a newspaper about the attempt to contain yellow fever in Brazil: In a laboratory, they also bred the very species of mosquito that spreads yellow fever. The female pupae were separated and destroyed, while the remaining male mosquitoes were exposed to X-rays for 42 minutes to damage their sperm so that they became incapable of procreation. The sterile males were then released in large numbers and mixed with the free-living mosquitoes. In this way, the mosquito population was greatly reduced, since the sterile males could successfully mate with the free-living female mosquitoes, but the brood then died. As an innovation, the laboratory then purchased a cobalt irradiator to replace the X-ray machine. The cobalt-60 source emits ionizing gamma radiation that takes only a little more than a minute to render the males infertile.

So in the above example, gamma radiation is about 40 times more harmful than X-rays. And in the Van Allen radiation belt, it has gamma radiation as described. Cobalt-60 sources are also used in the space industry to test electronic components for space suitability. Electronics are much less sensitive to ionizing radiation than humans.

In addition to protons and electrons in the Van Allen radiation belt, particle radiation from the Sun is added in space. The Sun constantly emits electrons, protons, and other charged particles as solar wind. During solar eruptions, also called solar storms, it emits huge amounts of charged particles combined with X-rays and gamma rays: the solar wind becomes a solar storm, with particles flying towards the Earth at speeds up to 800 km/s [44]. So it takes only two to three days for the storm to reach the Earth. A solar storm cannot be predicted even today, as I have heard this from two solar researchers and as NASA also writes [45]. With Apollo one had accepted thus the risk of perishing with a solar storm and finally however had luck. One had multiple luck, because 1969 was a year of a solar maximum, where the chance for high solar activity and solar storms is particularly high. Nowadays one does not want to take this risk anymore and therefore postpones manned Moon and Mars flights permanently for 10 to 15 years.

Solar storms are strongly attenuated on Earth by the terrestrial magnetic field. They can produce northern lights, which become visible up to the latitudes of Central Europe. However, disturbances have also occurred. For example, on March 13, 1989, a solar storm caused a power outage throughout the province of Quebec, Canada. [46] The strongest documented solar storm took place at the beginning of September 1859 and is called the Carrington event after the British astronomer; auroras occurred as far as the tropical latitudes of Cuba and Hawaii. [47]

To reduce the exposure of astronauts to cosmic rays, they should avoid or quickly pass through areas of high radiation so that they receive the lowest possible dose. In addition, the astronauts can also be shielded. You are probably familiar with this from X-rays at the dentist's office, where you are each covered with a lead vest during the exposure. The more material you have in front of you, the safer it is. The astronauts in the Command Module were not particularly protected. The structure of the Command Module is like a honeycomb sandwiched between aluminum sheets. This structure is also known as a sandwich structure and is often used in space travel: high stability is achieved with low weight. [34][32] Overall, the shielding effect of the Command Module is equivalent to that of a 7 mm thick aluminum shell. [48] If NASA now writes after Apollo that the radiation was never an operational problem [45] then this can be correct only in retrospect. If one considers newer analyses of NASA, there a clearly larger radiation protection is demanded than this was the case with the Apollo missions: in order to be armed against solar eruptions, a shielding with a 10 cm thick water cover is suggested. [49]

In other words, the radiation risk was downplayed or not communicated at all. The radiation protection in the Command Module was too low to sufficiently shield the radiation that should have been expected in 1969.

Only with luck, because the Sun was quiet during the flight, the astronauts could survive their Moon flight in the solar maximum year 1969 with a low radiation dose. [50]

Do you remember, during the fictitious Luna 1 flight, which was described in the Zürichsee-Zeitung in 1964, the successful penetration of the Van Allen radiation belt was still an important milestone. Later, one heard less and less about cosmic rays in the reports, and with Apollo they were no longer an issue at all. After the flight of Apollo 8 at the end of 1968, the same newspaper wrote that possibly the Apollo 8 team would now also be selected for the first Moon landing in June 1969. Nobody thought about the possibility that this team had already returned from its flight damaged by radiation. The climax to play down the risk of a flight around the Moon had been set by the director of NASA in December 1968 by equating it with the risk of a test pilot during the testing of a new type of aircraft. To this it must be said that in a lunar flight there are many other dangers in addition to the radiation risk. In addition, a spacecraft does not have an ejection seat, for example, with which one could save oneself in an emergency.

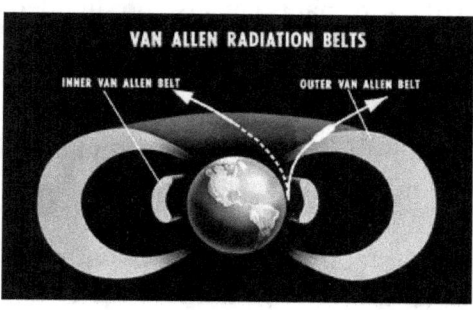

Fig. 21 Van Allen Radiation Belt with Flight Path

Although the shielding of the astronauts in the Command Module was low, the flight path through the Van Allen belt is well chosen. The Apollo 11 flight path pierces this belt at a narrow point. Since radiation was no longer an issue, NASA never communicated this to my knowledge. I reconstructed the flight trajectory myself using the description in the Mission Report [50]. Fig. 21 shows the Van Allen belt as well as the flight path in perspective by the initial dashed line with the arrow to the left. The line is extended once the flight path exits the Van Allen belt. The line with the arrow pointing to the right shows the penetration through the Van Allen belt in cross-section of the belt: the narrow penetration is clearly visible here. Surprisingly, the turning maneuver just described was performed while still inside the radiation belt, i.e., where the radiation was most intense throughout the mission. This location is shown in Fig. 21 thickly drawn.

I visited once a lecture about the Moon flight. The lecturer told that the penetration of the Van Allen belt happened according to the motto "Close your eyes and go through!" It was definitely not like that.

4.2 Swinging into Lunar Orbit

After one day of weightlessness on the way to the Moon, Apollo has just covered about half the distance to the Moon, the first of four planned course corrections is carried out: the rocket motor of the Service Module fires for three seconds to bring the Apollo spacecraft back exactly on course. This correction was prompted by tracking data from Earth. The astronauts have covered half the distance to the Moon, but in terms of time, this is only one-third. That means they will be two more days on the road before they can brake into lunar orbit.

Apollo 11 - THE Mission

The trajectory is then so precise that the three further planned course corrections are unnecessary. After 90% of the distance, the astronauts cross the point where the gravitational pull of the Earth and that of the Moon cancel each other out. Until then, their speed was getting slower and slower. Now that they are in the gravitational pull of the Moon, they speed up again. But the astronauts do not notice anything of it, they are weightless. One can compare this with a parabolic flight or with a suborbital flight: also here the passengers are weightless during the whole flight phase, thus with the ascending branch, where the speed decreases, with the apex and also again with the descending branch, where the speed increases again. This is exactly what happens to the astronauts on their flight to the Moon.

The spacecraft has rotated for braking into lunar orbit so that the rocket motor points in the direction of flight. After three days of flight to the Moon, the rocket motor fires during 6 minutes to turn into an elliptical lunar orbit from 69 to 190 miles (111 x 306 km) for the time being. This maneuver takes place behind the Moon, so there is no radio communication with the Earth anymore. From the elliptical orbit, after about two lunar orbits, a short firing of the engine for 17 seconds will then park into an almost circular orbit of 62 to 70.5 miles (100 x 113 km). [21][43]

The mileage figures vary in the documentation, as sometimes land miles of 1.609 km or sea miles of 1.852 km are used. In the Mission Report [21] the miles are uniformly in nautical miles. It also says that a circular park orbit with an orbital altitude of 60 nautical miles or 111 km was targeted. As described above, Apollo hit this orbit very quickly and quite accurately. The associated velocity is 1.63 km/s and the orbital period is pretty much two hours.

The rocket engines had to fire for only 17 seconds when braking from elliptical to circular orbit. This is because velocities in orbits between 100 and 300 km altitude above the lunar surface are almost the same: in circular orbits, the velocities at 100 km altitude are 1.63 km/s and at 300 km they are 1.55 km/s, only 80 m/s or 5% less, even though the orbit is, after all, three times higher. "Three times higher" is correct, but for the calculation of the velocity the radius of the orbit is decisive, i.e. one must add the Moon radius of 1,738 km to the orbit height. The radii of the orbits are then 1,738 km + 100 km = 1,838 km respectively 1,738 km + 300 km = 2,038 km and finally differ only about 10%, so it is not surprising that the corresponding velocities are also close to each other.

Apollo 11 – The Real Story

Some of the astronauts' activities are well documented thanks to the Hasselblad cameras they carried. These are single-lens reflex cameras of the type 500EL. EL stands for electric. This is not to be confused with the electronic cameras with digital image sensor that are common today. Electric here means that after each exposure, the motor advances the film by one frame and cocks the shutter: the camera is then immediately ready for the next exposure. The films are exposed when the picture is taken and only later developed in a liquid in the laboratory, so that an image is formed on the film. The images can be projected as slides [51] onto a screen; here we look at the scanned versions.

At the back of the cameras is a cube-shaped magazine containing the 70 mm wide film. The image format is square 54 mm x 54 mm. [52]

The films are numbered consecutively per magazine. One magazine holds one roll for 160 color images or 200 black and white images. I present here first photos from magazine number 37, with which the color images 5433-5555 were taken, a total of 123 images. These and all other images are described and available in the Apollo 11 Image Library [53] and can be viewed. The image designations are AS11-37-5433 through AS11-37-5555. These images were taken indoors in the Lunar Module, first from lunar orbit and the last images after landing. An f-2.8/80mm lens was used for these images [52].

Like the images still to follow, these images are labeled "OF300," which are scans from the original film at a resolution of 300 dots per inch on an image size of 7.5 x 7.5 inches, which is 2,250 x 2,250 pixels.

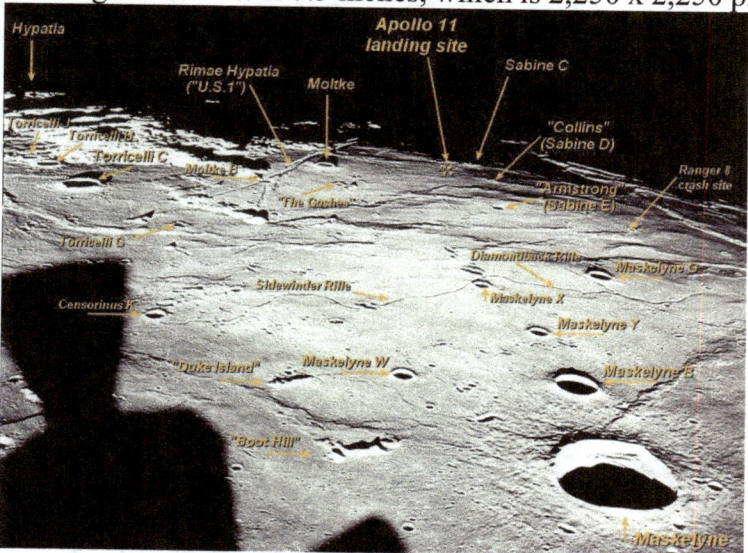

Fig. 22 AS11-37-5437 labeled as first General Map

Apollo 11 - THE Mission

Three hours or one and a half lunar orbits after entering the circular orbit - Neil Armstrong and Buzz Aldrin have already transferred to the Lunar Landing Module and are about to activate it - Buzz Aldrin glimpses through a window the landing site looming ahead and takes a photo (Fig. 22). The landing site is labeled "Apollo 11 landing site". The day-night boundary lies 0.6° [53] behind respectively west of the landing site, accordingly the Sun is low on the horizon seen from the landing site. The day-night boundary runs from north (right in the picture) to south. It will take about one more day until the landing and the exit, so that the Sun will be higher by then.

The nozzle on the lower left belongs to the attitude control system of the Lunar Landing Module.

4.3 Separation and Landing Maneuvers

Apollo has now been at the Moon for one day, has swung from the elliptical to the circular orbit four hours after arrival and has orbited the Moon ten times. The astronauts have prepared everything for the following maneuvers and are also rested. The next step is for the Lunar Module to detach from the Command and Service Module CSM so that it can then fly from its current 111 km circular orbit into a lower orbit and then land. In the Lunar Module are now Armstrong and Aldrin, while Collins alone tends the CSM.

First, the astronauts perform the undocking maneuver: The CSM and Lunar Module undock and slowly drift away from each other. Once the distance has grown to 12 m, Collins brakes the CSM so that the two spacecrafts are now traveling at a constant distance. The Lunar Module then makes a full horizontal rotation (i.e. around the axis pointing to the Moon) so that Collins can see whether everything is OK from the outside. Then the two spacecraft continue to fly in this formation, with their distance changing by less than two meters per minute.

I did not find in the documentation in which direction this undocking went. Flying side by side would not have been a good idea, as this only works for a short time in space. Imagine two satellites orbiting the Earth from north to south and then again from south to north, thus flying over the North Pole and the South Pole again and again. The trajectories then follow exactly a longitude line each, if we consider the Earth to be stationary for this thought experiment, i.e. not rotating. If now these two satellites fly over the equator side by side in the direction of the South Pole, satellite 1 on the left and satellite 2 on the right, then they collide over the South Pole, since all longitude lines converge there. If there is no collision, for example because one satellite is slightly behind, then satellite 1 flies to the right and satellite 2 to the left, until it changes again at the North Pole. But back to the undocking at the Moon: probably the CSM did not drift away vertically upwards or downwards, otherwise Collins would have seen only the same side during the rotation of the Lunar Landing Module. So I suspect that the CSM and the Lunar Landing Module separated in flight direction and then flew behind each other. But I will try to clarify this point still on the basis of the photos.

In principle, the Lunar Module could now fly directly from this constellation to a lower orbit and initiate the landing. But to prevent the CSM from being hit by the return beam of the Lunar Landing Module or to prevent a collision in case of an error, an actual separation maneuver follows first, during which the two spacecraft move to a distance of several hundred meters. During this separation maneuver, it is the Command and Service Module CSM that uses its attitude control thrusters to slowly accelerate downward, toward the Moon, until it moves away from the Lunar Landing Module at a speed of 2.5 feet/s or 76 cm/s. [21]

I was surprised to see that now the CSM flies down, where actually the Lunar Module should take the way down. But it also works like this, as we will see shortly, and the two astronauts in the Lunar Module can still take photos of the CSM flying away with the lunar surface as background.

For the separation maneuver I found a nice diagram in the Press Kit [28], an overview document of NASA, which is reproduced in Fig. 23. We look from "above", i.e. from above the lunar North Pole on the Moon, and we see up to the lunar equator. The landing site (LANDING SITE) is drawn on the equator, where it lies with 0.67°N approximately. The Sun has just risen there. The night side of the Moon is left of the thick black line and colored gray. The hatched segment indicates in which angle also the space ships are in the dark. Since these fly 111 km above the lunar surface, they dip only later into the Moon shadow. You can observe this phenomenon also with us: if you look at a satellite after sunset, you see it exactly because it flies high and is still illuminated by the Sun. The satellite can then suddenly disappear after a while, namely when it enters the Earth's shadow.

The two large circles outside indicate the orbits of the two spacecraft: extended, the original circular orbit with an orbital altitude of 60

Apollo 11 - THE Mission

nautical miles or 111 km. This is the orbit in which the CSM and the Lunar Module flew when they were still coupled together, and they remained in this orbit during decoupling (point 1). The CSM now accelerates toward the Moon during separation (point 2) and continues to fly on the dashed orbit, while the Lunar Landing Module remains on the extended circular orbit. It continues to fly there for half a lunar orbit until it begins its descent to the Moon at point 3. But for now we stay with the separation maneuver and the resulting trajectory. The orbit altitude is drawn very exaggerated, also the difference of the two orbits is shown enlarged, so that one can see the sequence and the different trajectories well.

You can see that the two orbits intersect again 180° after separation. Are the two spacecraft on a collision course? They are not with respect to point 3, as another diagram in the Press Kit shows (see Appendix, §10.4). The Command and Service Module CSM will pass there 1.8 miles or 3.3 km ahead of the Lunar Landing Module, allowing it to brake in for descent undisturbed. But if for some reason this braking-in maneuver were delayed by one lunar revolution, the two

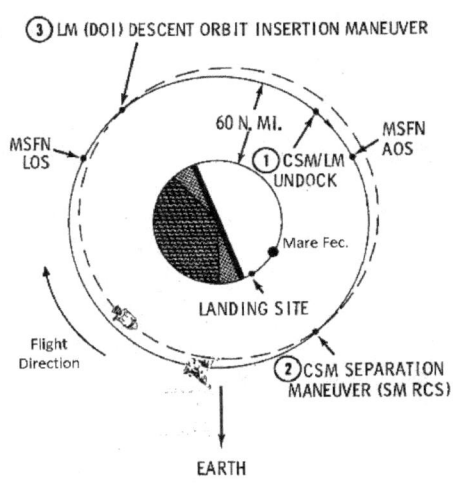

Fig. 23 Separation Maneuver [28]

spacecraft would again come very close to each other at the original separation point (point 2). They would then have to be careful not to collide. Then the Lunar Module could try a second time to initiate the landing. But it did not come so far, everything went according to plan.

As with the example where two satellites fly side by side, one does not immediately understand why the CSM arrives at point 3 before the Lunar Module. But if you look Fig. 23 closely, you see that the CSM takes a shortcut from point 2 to point 3 and therefore reaches point 3 before the Lunar Module. For this, the CSM has to make a detour on the second half of its orbit, so it would lose its lead and then arrive again almost simultaneously with the Lunar Module at point 2, if the Lunar Module had not initiated the descent in the meantime. The description of this maneuver is therefore consistent.

Two things in Fig. 23 I have additionally inserted: on the one hand the arrow of the direction of flight and on the other hand the point which I have written "Mare Fec". I will come to this in a moment.

By the way, the two dots "MSFN LOS" and "MSFN AOS" indicate the places where the radio contact to the Earth breaks off (left) and from where it works again, respectively.

As mentioned earlier, the two astronauts in the Lunar Module took the opportunity to take pictures of the CSM and the Moon below. I present three of these images here. In the first two pictures the CSM is still quite close and well visible (Fig. 24 and Fig. 25):

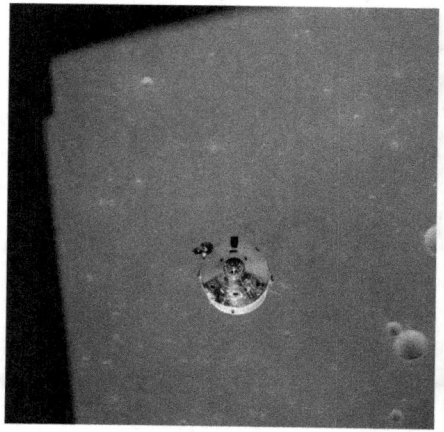
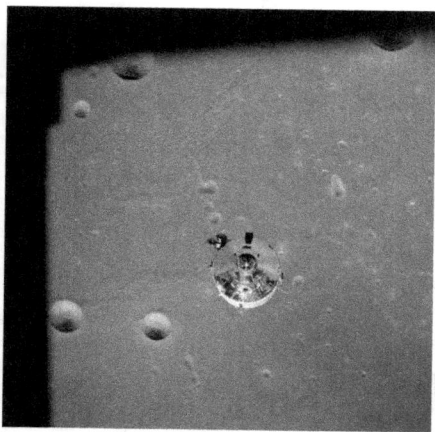

Fig. 24 CSM as seen from the Lunar Landing Module (AS11-37-5444). [53][54]

Fig. 25 Once again the CSM seen from the Lunar Landing Module (AS11-37-5445). [53][54]

In the description of these pictures is indicated that just the Mare Fecunditatis is flown over and likewise the names of some craters are indicated there. The view to the lunar surface is steep downward and slightly forward; one looks approximately in flight direction. The CSM is almost exactly the same size on both images, so it is also the same distance away. This fits to Fig. 23 according to which the two spacecraft are still traveling at a constant distance above Mare Fecunditatis. So during the undocking maneuver the CSM drifted away rather downward and not forward or backward as I suspected. During the inspection of the Lunar Module by Michael Collins from the CSM he would have seen then always the same side of the Lunar Module, because the Lunar Module has turned around its vertical axis. But maybe the CSM drifted away diagonally downwards, so that Collins could look at the LM from different angles.

From the two images, one can estimate the distance between LM and CSM: knowing the diameter of the CSM (3.9 m), one only needs the angle at which one sees the CSM. In the data sheet of the lens [55] is written that the lateral aperture angle is 38°. Since these images are slightly taller than they are wide, I took 37° as the image width and read out the angle that the CSM makes directly relative to the image width in centimeters. A similar method of distance measurement is commonly known as thumb jump: if you aim at a distant object with your arm outstretched over your thumb with one eye and then with the other eye, the thumb makes a lateral jump. If you know or can estimate the lateral distance at the distant object, then multiply that by 10 to get the approximate distance to the object. This works perfectly for me because my two eyes are 6 cm apart and the outstretched thumb is 60 cm in front of my eyes.

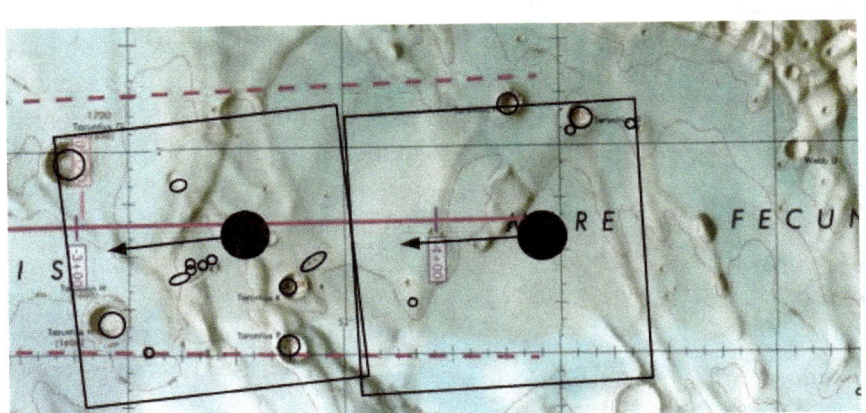

Fig. 26 Approach Map with the Breaks of the Images of Fig. 24 and Fig. 25

So I measured the angle at which you can see the CSM in both images and calculated the distance: in the first image the result is 37.6 m and in the second one 38.4 m. So the CSM is almost the same distance away in both images as expected and the distance has increased from the initial 12 m to about 38 m, which fits to this flight constellation, because the orbit altitudes are slightly different and the distance was not kept actively constant.

To show where these images were taken, I created a pause of these images, transferring the most important contours, i.e., image edge, crater, and the CSM to the pause. From the CSM, I drew an arrow to the top of the image as a guide. I transferred these two pauses to a Moon map and aligned them as closely as possible with the map by rotating and adjusting the size (Fig. 26). I changed the size only as a whole, so I kept the aspect ratios of the images; circles remain circles and do not become ellipses. Many craters on the pictures are also circular, because one looks steeply downward. As Moon map I use the so-called approach map [56] on which the approach corridor of the Lunar Landing Module is drawn. The craters, which one sees according to the description of NASA on the pictures, agree with my placement of the pictures.

On the lunar map, longitude and latitude are drawn in a two-degree grid. Two degrees on the Moon - at the equator, longitude and latitude are of equal length - correspond to 61 km.

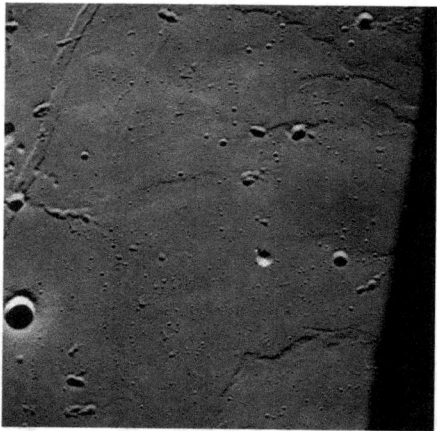

Fig. 27 CSM and below the Landing Site photographed from LM (AS11-37-5447)

Above the landing site, the astronauts took a third picture (Fig. 27).

The CSM is now much farther away. The distance has grown to 167 m due to the separation maneuver. I also took a break from this image and transferred it to the lunar map. (Fig. 28)

Fig. 28 shows that the timing for this image (AS11-37-5447) could hardly have been more perfect: the CSM touches the edge of the intended landing site. In a little more than two hours, the Lunar Module will land at the coordinates 0.67°N, 23.47°E [57]. I have drawn the location as a dashed line on the map. By the way, on the lunar map the equidistance of the elevation curves is 600 m and all elevations are in meters. The number 1070 at the bottom of crater Moltke in brackets indicates the depth of the crater measured from the upper crater rim and 470 is the absolute height of the crater rim.

For the image of the landing area (Fig. 27) the angle of incidence of the Sun is about 10°, as I estimated from the mission times. One recognizes very nicely the sharp shadows of the craters. Since there is no atmosphere on the Moon, shadows are generally sharply drawn.

What surprised me, however, is that the break-off edge around crater Moltke (at the left edge of the image, below) is not visible, despite the flat sunlight. In the two images above Mare Fecunditatis, the Sun is 28° higher, so about 38°. Here I would have expected just as sharp shadows as on the picture above the landing site.

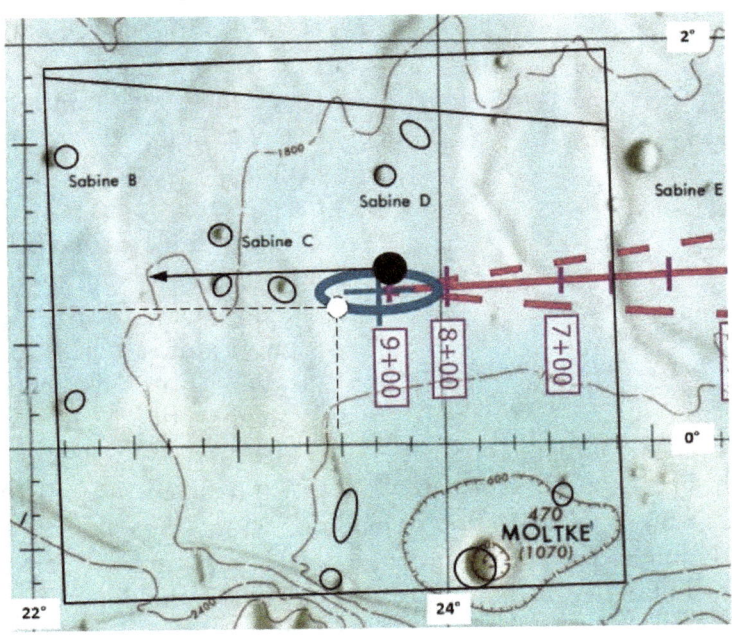

Fig. 28 Approach Map with the Break from the Image AS11-37-5447

We now return to the upcoming landing: after the Lunar Module has passed over the landing site, the two astronauts in the Lunar Landing Module are expected to begin the landing maneuver in just under an hour. This action is described in Fig. 23 as Descent Orbit Insertion (DOI). Fig. 29 shows the descent trajectory. For now, it is sufficient to decelerate only slightly in order to enter an elliptical orbit and thus get closer to the Moon. As I mentioned in the last chapter, the orbital velocities at low orbits are all similar in size. In this case, the velocity of 1.63 km/s only needs to be reduced by 22 m/s to enter the desired elliptical orbit of 60 miles x 50,000 feet, or 111 km x 15 km. So during the next half orbit of the Moon, the Lunar Module will fly down to an altitude of 15 km, taking about an hour. After this free flight, the phase now begins in which the engine fires permanently, thus continuously reducing the speed to zero. The beginning of this phase is called Powered Descent Initiation (PDI). The landing site is at dawn, the Sun is still quite low. The approach comes from the sunny side, the Sun shines on the Lunar Module from behind.

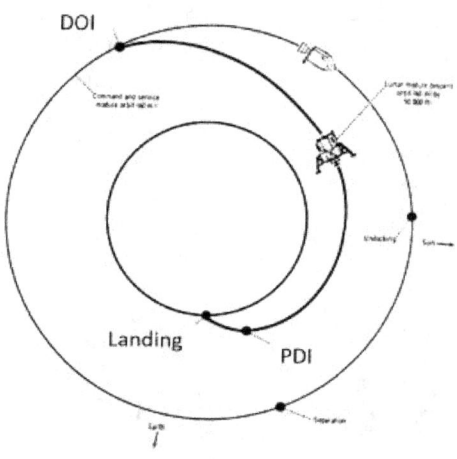

Fig. 29 Descent Trajectory of the Lunar Landing Module [21]

With PDI, the trajectory is horizontal, as can be seen in Fig. 30. The Lunar Module decelerates in this phase with about 1/3 of the acceleration due to gravity on Earth, and the two astronauts feel 1/3 g. So it is reasonable to expect them to stand. This first part of the deceleration phase of Fig. 30 lasts 8 ½ minutes and is between 260 miles to 4 miles from the landing site. Since initially the flight is almost horizontal, some of the thrust is needed to maintain altitude. Although the main part of the thrust is opposite to the direction of flight and reduces speed, a smaller part points upward and is needed only to maintain altitude. This method allows landing with a weak engine and correspondingly low acceleration, so it is well suited for manned landings, but it requires more energy. Today's landings, where the rocket engine is used for braking, are performed steeply from above, as we will see in a moment.

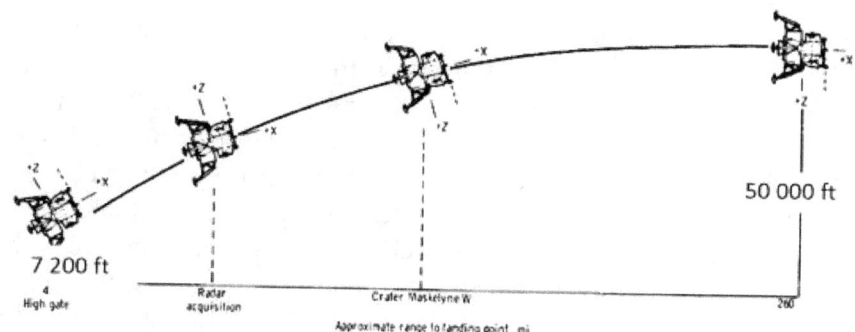

Fig. 30 Landing Approach between an Altitude of 50,000 and 7,200 Feet [21]

The final phase or the last 5,000 feet (1,500 m, horizontal) to the landing site is shown in Fig. 31. The initial altitude is 1,000 feet or 300 m. The approach is still flat; on Fig. 31 it is shown elevated.

Apollo 11 - THE Mission

Although the shallow approach allows the crew to continue to a suitable landing site in the event of unsuitable landing conditions, the Lunar Module requires permanent upward thrust.

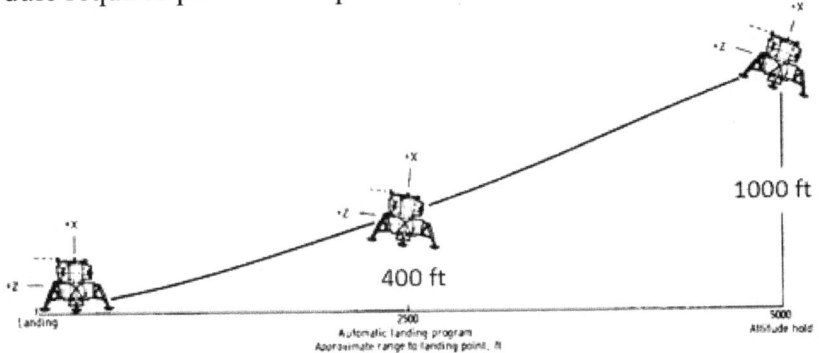

Fig. 31 Final Phase of the Landing [21]

The engines burn for a total of only 12 ½ minutes in the landing phase, until the Lunar Landing Module then touches down on the Moon on Sunday, July 20, 1969 at 20:18 Universal Time, in Central Europe it is 21:18, and Neil Armstrong can report: "The Eagle has landed."

Did this landing take place as just described or is NASA telling us a made-up story here? To judge this, we can compare this landing with similar maneuvers. For this purpose, the training results in the run-up to Apollo 11 as well as today's rocket landings are suitable. Let's start with the training results.

In the chapter 3.2 I introduced the training flight and landing device LLRV, also called flying bed frame (Fig. 7). With this device the astronauts practiced flying with the Lunar Module. In the LLRV, the astronaut sat on an open cockpit at the edge and had a good view from there. The similarity to the Lunar Module is slight: in the Lunar Module the astronauts stand in the cockpit and they can only look out through small windows. In addition, the astronauts only practiced hovering and slowly moving back and forth above the ground with the LLRV. Deceleration from high speed could not be practiced with the LLRV. During the Moon landing they flew in vacuum, at high speed and under the influence of the low gravity of the Moon. With this nevertheless poor correspondence from the test experiment to the serious mission, I would expect the test experiment to show that the control of these devices is very well mastered; that is, that one can accept the additional uncertainty not covered by the test experiment. However, it was exactly the opposite: the astronauts did not master the test device in any way, and NASA itself called Armstrong's flight a *Rehearsal Mishap* (Fig. 8). Therefore it seems to me very questionable that everything went like clockwork at the Moon despite failed tests.

We can also compare the landing of the Lunar Module with today's state of the art. The engine during the landing was the rocket engine at the lower end of the Lunar Landing Module, at most with the support of attitude control thrusters. Similar landing attempts have been made on Earth in recent years with rocket stages. The rocket stage, like the Lunar Module, has one or more rocket motors that are used to maintain balance and are needed for braking. Such landing attempts, using a rocket motor at the bottom of a rocket stage to try to land, have been seen repeatedly in recent years. Then, after many failed attempts, on Monday, Nov. 23, 2015, in Texas, the space company Blue Origin made history: for the first time, a rocket was able to land softly on Earth by using its engine to brake and stabilize.[58] It touched down in an upright position at a speed of about 7 km/h (Fig. 32). Shortly before, it had launched at the same location and ascended to an altitude of 100 km, where it had released an unmanned passenger capsule. [59]

Fig. 32 November 23, 2015: First soft Landing of a Rocket Stage on Earth [60]

On Monday, December 21, 2015, the landing of a rocket stage also succeeded for the first time for the SpaceX company; this time even on a commercial flight, releasing 11 ORBCOMM 2 satellites into space. The lower stage of a Falcon 9 rocket landed back at its origin at Cape Canaveral in Florida (Fig. 33). [61]

Fig. 33 Takeoff and Landing of the Falcon 9, recorded with a long time Exposure. [61]

Both landings were steep from above. The rocket stages were longer than the Lunar Landing Module, so balancing on the lower rocket engine was more difficult, and in the 6 times stronger Earth gravity field one has to correct faster, but the rocket stage flew in air which helped to stabilize the attitude. Besides, the attitude sensors are better today than in the 60's and also the rocket motors can be controlled more precisely.

Apollo 11 - THE Mission

All in all, the described landing on the Moon fits poorly with the development of landing techniques: before 1969, it was nowhere near possible to land a rocket in front of witnesses, and afterwards it took until 2015, 46 years, for this to become possible.

The fact that landing on the Moon is difficult to control is also supported by the fact that after Luna 24 in 1976, there was not a single soft landing on the Moon until December 2013, when China placed its Jade Rabbit on the Moon. [62]

4.4 Moonwalks

4.4.1 General Information about the Moonwalks

A little more than 6 hours after touchdown on the Moon, the outdoor mission begins at 02:40 Universal Time on Monday, July 21, 1969, with the opening of the door of the Lunar Landing Module. Before Armstrong climbs down the ladder, he brings the TV camera into position. To do this, standing at the top of the ladder, he has to pull on a locking ring so that the container in which the TV camera is mounted swings down out of the side wall of the Lunar Landing Module into its working position like a folding bed or murphy bed. I will come back to the mounting of the TV camera. About a quarter of an hour after the opening of the door Neil Armstrong then speaks the famous sentence „That's one small step for a man, one giant leap for mankind." when stepping on the lunar surface. [218] So he says this sentence around three o'clock world time; in Central Europe it is already almost four o'clock in the morning, but on the American East Coast it is still 11 o'clock on Sunday, July 20, 1969. For this reason sometimes the 20th July and sometimes the 21st July is mentioned as date of the stepping on the Moon. 18 minutes after Armstrong, Aldrin also descends the ladder. Armstrong and Aldrin's exits are broadcast live to Earth.[63] This is the magic moment when half of Europe and America are in front of the

Fig. 34 Legendary Footprint (AS11-40-5877)

television. Most of those who were there still remember this moment today.

All time data are described in the Apollo 11 Lunar Surface Journal [64]. The images can be found from there by clicking the "Image Library" link and the videos by clicking the "Video and Movies" link. There are also descriptions and transcripts of conversations which I sometimes quote here.

The time from opening the door to closing it again is a little more than 2 ½ hours. This phase is called the extravehicular activity (EVA). During this time, the astronauts get out and back in at the end, in between taking photos, collecting lunar rocks, and setting up the solar wind sail, flag and equipment for further experiments. I will discuss the different experiments in the chapter 7 in detail.

For the general public, probably the most important things they do are the photographs and film footage. It is mainly the photos that are still seen today and some of them are symbolic, for example the legendary footprint (Fig. 34) or the astronaut next to the flag (Fig. 40). It is mostly Neil Armstrong who takes the pictures. That's why you almost always see only Buzz Aldrin in the photos. There is only one picture of Neil Armstrong, namely the picture with the designation AS11-40-5886 (Fig. 35). AS stands for Apollo-Saturn, 40 is the number of the magazine and at the end follows a consecutive number.

Fig. 35 The Only Picture of Neil Armstrong (AS11-40-5886)

As with the images taken from inside the Lunar Landing Module, a Hasselblad SLR (single lens reflex) camera is used on the lunar surface. Its designation is "500 EL Data Camera". The Data Camera is also a modification of the standard 500 EL camera, but differs in the following ways:

a) The camera is equipped with a reseau plate. This is a glass plate on which crosshairs are engraved. This glass plate is close in front of the film, so that black crosses are visible on the finished image. The cross in the center is larger than the others. The distance between two crosses is 10.3°. If there had been such crosses on the pictures of the CSM and the lunar landscape below, the evaluation would have been a little easier for me.

b) The data camera is equipped with a Biogon f-5.6/60mm lens. The focal length of 60 mm is slightly smaller than the camera used in the Lunar Module. Accordingly, the (lateral) recording angle is larger at 47°. [65]

c) Instead of black, the data camera including the magazines is silvery on the outside so that it heats up less in the Sun (Fig. 38). Everything on the Moon that is illuminated by the Sun heats up strongly because there is no air to transport the heat away again.

d) The camera is designed so that it does not become statically charged. I'm sure you've also had the experience of getting a static charge when walking across plastic carpets in dry air, and then getting electrified as soon as you touched a doorknob or metal rack. Vacuum is worse than dry air in this regard, so surfaces of satellites and their measuring devices are each coated with electrically conductive materials. NASA was thus able to prevent the film from being damaged by electrical discharges.

Armstrong has a mount on his chest to which he attaches the camera. He is unable to look through the viewfinder, but aligns himself with the object to be photographed and presses the shutter release button, which is mounted at the front in such a way that this works with the gloves on. In Fig. 37, Aldrin has his finger straight on the shutter button. When the astronaut is standing upright, the camera is straight or well horizontal and the line of sight is horizontal. This means that vertical lines are also largely vertical in the image; standing people, for example, are shown upright and not at an angle. Today, there are cameras that use a tilt sensor in the viewfinder to indicate whether you are holding the camera straight. If the astronaut leans forward or backward, then the camera is tilted down or up. In this case, it remains horizontal and only the vertical center line is displayed vertically; lateral vertical lines are displayed in perspective. (Fig. 36)

Fig. 36 View down with Camera well levelled

Distance, aperture and shutter speed can be adjusted on the rings, but this is difficult to handle with gloves, so these values are generally fixed on the Moon, which suits most shots. The image of the footprint (Fig. 34) was taken by Aldrin holding the camera in his hands. Here he must have adjusted the focus ring, because in this picture the near part is in focus and the more distant part is out of focus. In the landscape shots, the distant objects are in focus in each case. The Hasselblad camera was *the* professional camera par excellence in the 1960s. In contrast to today, all camera settings were made by hand: focus, shutter speed and aperture were set here on rings. The camera is electric, but "electric" refers only to the automatic film transport and the tensioning of the mechanical shutter. This camera has no built-in light meter and therefore no automatic exposure control. To determine exposure, photographers used a separate light meter.

Fig. 37 Position of the Camera in Training (ap11-)S69-31109 (detail) [53]

Fig. 38 Aldrin with the silvery Data Camera in Training (KSC-69PC-362) (detail). [53]

In the photos where the camera was in Armstrong's chest mount, I estimate its height above ground to be between 1.35 and 1.50 meters. Armstrong was 180 cm tall [28]; by the way, he was the same size as me, so I can measure this height directly on me. When looking closely at some photos, it is good to know the camera height.

Apollo 11 - THE Mission

The two astronauts take a total of 122 pictures with the magazine 40 with the numbers 5850 to 5970 plus the picture 5882a. The time from the first to the last picture is 1 hour 56 minutes and 12 seconds, which means that on average only 58 seconds elapse between two pictures. So that the photographers among you do not become too jealous of this achievement, I can add that four 360° panoramas and one 180° panorama à 8 to 12 photos with a total of 53 pictures were taken [66] and that Armstrong pulled the trigger 6 times quickly at the end to wind the film forward before changing the magazine, in order to protect the good lunar images in case of light entering the magazine. Altogether there are 59 photos, which were taken shortly after each other; for the remaining 63 pictures then only about a 2-minute cycle results.

The angle of solar incidence is fairly constant during the 2½ hours of this outdoor mission - after all, one day on the Moon lasts 29½ Earth days. The angle of incidence is 14° at the beginning and 15.4° at the end. [67]. This corresponds to the altitude of the Sun in Zurich on March 21, i.e. at equinox and still in winter time, at 8 o'clock in the morning. One has chosen the early morning as landing time, because at this time the lunar soil is still cold. The temperature of the lunar soil drops during the lunar night at the equator down to -173°C and reaches its maximum of 107°C at noon. At the time of the outdoor sojourn, the lunar surface temperature is -23°C. [68] So the cold ground is a welcome compensation to the intense radiation of the Sun.

The second camera used on the Moon is the TV camera (chapter 4.4.3, Fig. 62). It is used, among other things, for the live transmission of the exit of the two astronauts. For this scene, the wide-angle lens is used, which is intended for indoor shots and scenes with indirect solar illumination. It has an image diagonal of 80°.[69] That actually this lens was used I checked by the size of the upper part of Aldrin's body in the picture. You will find the estimation in appendix §10.6.

4.4.2 Shape of the Landscape around the Landing Site

Looking at the topography of the landing site first may seem strange to you. We are conditioned in this subject to admire the flag, the footprint, the saluting astronaut or other details. I would like to give us the overview here first and only move on to the details afterwards.

The landing of Apollo 11 took place in the Sea of Tranquility (Mare Tranquillitatis). The seas are the dark areas of the Moon. They are thought of as solidified lava basins that were filled in primeval times by lava flowing out from the lunar interior. As the name implies, they are quite flat. Also if we look at the lunar map (Fig. 28), we see that the terrain is not perfectly flat, but it is not mountainous. How it looks exactly at the landing site cannot be read from the map, for this the scale is too small and the elevation curves too rough.

In the photos around the landing site, neither mountains nor hills can be seen. The terrain appears flat, which fits perfectly with the image of a sea (Fig. 39 and Fig. 40).

Fig. 39 Aldrin setting up the Solar Sail (AS11-40-5872)

Fig. 40 Aldrin at the Flag (AS11-40-5875)

In the picture with the solar sail, the upper edge of Aldrin's helmet touches the horizon. Aldrin is 178 cm tall [28] thus only two centimeters shorter than Armstrong, who took the picture from a height above ground of 135 to 150 cm. So the horizon is higher than the landing site and rises again towards the left part of the picture. In the picture with the flag, however, the horizon is perfectly horizontal and lies at the height of the camera and is also exactly at chest height for Aldrin. The terrain in the background therefore appears in this section of the image in the direction of view and perpendicular to it horizontal.

The visibility to the horizon in the image with the flag is 2.2 to 2.3 km on level terrain (the calculation is given in Appendix §10.2). On the Earth with its larger diameter, for example, the visibility at the sea from the same camera height would be 4.1 to 4.4 km to the farthest point on the sea surface that one can just see. Since one does not notice its spherical shape on Earth, this is obviously the same on the Moon. The horizon still appears very far away and horizontal with more than 2 km in the perfectly flat terrain.

Fig. 41 Aldrin on the Ladder (AS11-40-5868)

Fig. 42 Image Composition (as11-5864-69) [70]

Fig. 43 View to the northwest: Aldrin at the Lunar Landing Module (AS11-40-5928)

Fig. 44 View to the north: Aldrin at the Seismometer (AS11-40-5948)

In Fig. 41 the horizon rises to the right, from which one could conclude that the terrain also rises to the right. But in the composite image as11-5864-69 (Fig. 42), the same horizon line is much flatter; probably in Fig. 41 the camera was tilted sideways, thus poorly horizontal. In general, if there is no reference for the vertical, it is impossible to judge whether the camera was level and whether the terrain is sloped or flat. Therefore I present with Fig. 43 and Fig. 44 two images with a good reference for the vertical. In both images, you can see from the astronaut, the Lunar Module, and the flagpole, the lower part of which can also be seen on AS11-40-5928, that the images are well levelled.

The Sun is in the east and the shadows point to the west. So you look to the north in the picture with the seismometer. To the left of the Lunar Module its shadow is visible, which runs horizontally.

In the picture with Aldrin at the Lunar Module (Fig. 43) the shadow of the Lunar Module is visible in its whole length. The distance from the camera to the horizon is extremely short and can be estimated according to Fig. 45 to a maximum of 38 m.

Fig. 45 AS11-40-5928 with approximate (rounded up) Distance Data

The two lower arrows start from Armstrong's feet. These are directly below the camera and therefore lie in the extension of the vertical center line of the image.

The length of the arrow from Armstrong to Aldrin I estimate with the help of the height of Aldrin (1.8 m) and the angle, under which one sees him, to 8.8 m and round this for Fig. 45 to 10 m. The shadow of the Lunar Landing Module is 28 m long, which results from its height of 7 m and the Sun incidence angle of 14°. The 38 m from Armstrong to the end of the shadow touching the horizon results from the generous addition of the two lengths. Such a short distance to the horizon almost never occurs in daily life. It could occur, for example, when looking up to a nearby ridge on a mountain tour. On the Moon, such a short distance to the horizon would be possible in a crater, if one looks up to the rim. But since all these images point to a flat terrain, I would expect more lunar landscape in the background.

What I have heard to this picture, by the way, is that the visible horizon is in reality the day-night boundary on the Moon. This is by far not so. At the day-night boundary the Sun is at the opposite eastern horizon and the Sun incidence angle is 0°. The shadow boundary was in the first picture, which shows the landing site from the Lunar Module (Fig. 22), already 0.6 degrees of longitude west. In the image "Aldrin at the Lunar Module" (Fig. 45) the Sun incidence angle is (at least) 14°, so the light-shadow boundary is now 14 degrees of longitude or 425 km in the west.

Fig. 46 AS11-40-5928: Aldrin at the Lunar Module, with Horizon Lines

As a next step, I have drawn the estimated height of the camera on this image: this dashed horizontal line would correspond to the horizon in level terrain. This line is also called the mathematical horizon. In the picture with the flag the visible and the mathematical horizon are identical. In the picture "Aldrin at the Lunar Module" the extension of the visible horizon (dash-dot-line) on the right of the picture intersects the mathematical horizon in the vanishing point, which is also used in perspective drawing. The dash-dot line of the visible horizon fits a horizontal or planar course.

In the left half of Fig. 46 one looks clearly down to the horizon and also "down to space". The estimated angle with which one looks at the end of the shadow of the Lunar Module is calculated in flat terrain as 1.35..1.5 m height to 38 m length equal to 2 to 2.3°, which fits very well to the drawn angle of 2.5°, because I rounded up two times at the length. To 2.5° fits exactly 1.5 m height and 34 m distance. On the far left of the picture this effect is even more pronounced, because the distance to the horizon is shorter.

Looking down to space would only be possible on a lonely elevated platform. Only at 2.5° this platform would have to be 1,600 m above the Sea of Tranquility and there should not be a hill to be seen protruding above this edge for the next 76 km. This is shown in Fig. 47. The lunar map (Fig. 28) with elevation curves every 600 m shows that there are no such landforms there.

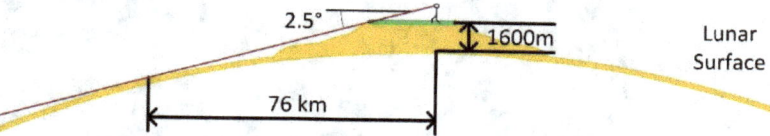

Fig. 47 Landing Plateau and Sighting Line over the Edge

If you closely look at Fig. 47 you can see that the near horizon described by me could also be dragged into further lunar background, so that you could see on AS11-40-5928 instead of the maximum 38 m up to 76 km far; at most even further, if there was another elevation behind it. Strictly speaking, it is not the short distance to the horizon which leads to the contradiction, but the fact that one looks so steeply down to the horizon.

I recreated the scene on a soccer field and could only get the same horizon progression if I covered the background (Fig. 48 and Fig. 49).

Apollo 11 - THE Mission

Fig. 48 AS11-40-5928

Fig. 49 Re-enacted Scene with a similarly deep Horizon Gradient - thanks to darkened Background

Above illustrations and considerations suggest that the horizon line marks the end of a studio scene and not a real Moon horizon. All pictures fit to each other; they were taken in the same environment. So if one picture had been taken in a studio, this would apply to all pictures - and of course also vice versa for a lunar landscape as shooting location.

It would be quite clear if it could be shown by a single picture whether the Lunar Module is in a studio or on the Moon. Because taken by itself the photo AS11-40-5928 is no proof for a studio photo. One can only judge together with the other photos that the Lunar Module is not standing on a slope and that one does not look upwards and see the upper edge of a hill as horizon. By the way, such a scenario would be supported by the statement of NASA that the Lunar Module is inclined by 4.5° to the east [21]. According to image AS11-40-5948 (Fig. 44), however, its shadow runs horizontally, so one must not conclude here from the indicated inclination of the Lunar Module to an inclined slope. For a road with a slope of 4.5°, the difference in height per 10 m length is 80 cm; if it rises by at least 1 m per 10 m length, then in Switzerland the danger signal "steep slope" is set up. [71] The terrain around the Lunar Module is far from such a gradient.

In the live television broadcast I found the desired single scene. Fig. 50 shows a still image of this video again together with a reenactment on the soccer field (Fig. 51). I indicated the approximate camera height with the dashed line. Aldrin is standing upright and you can see the top of his helmet and his backpack.

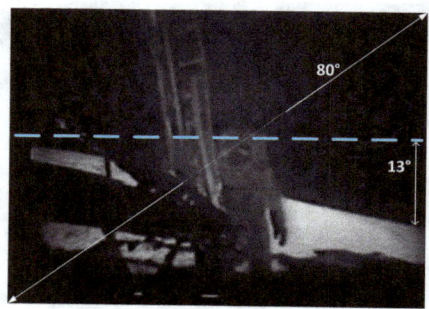 

Fig. 50 Still Image from the live TV Broadcast [72]

Fig. 51 Re-enactment with same Horizon Line thanks to covered Background

Fig. 52 Training Image (ap11-)S69-32247

At Fig. 50 the "looking down into space" effect is so obvious that this one image may be regarded as mathematical proof that this film shot could only have been made in a studio. Even if the camera were not exactly horizontal or were a bit lower than drawn, the "looking down into space" effect remains so strong that neither on the Moon nor on the Earth there is a natural landscape which makes such a shot possible. On the right side of the TV picture the horizon is 13° below the camera height (dashed line). This would require a plateau more than 45 km high on the Moon and no other high mountains within a radius of 400 km; on Earth, the conditions would be even more extreme. So in this example, the margin is so large that this video alone is sufficient to expose the shooting location as a studio on Earth: you can see the illuminated floor and behind it is a black wall suggesting outer space. In Fig. 51 the black wall corresponds to the darkening drawn in. I have found a training image (Fig. 52) where this effect of the deep horizon is so clear that one would immediately perceive the artificially illuminated ground as such even if the background were perfectly black. Here, Aldrin places the solar sail too

Apollo 11 - THE Mission

close to the horizon for this image to be sold as a true lunar image with some re-blackening.

If you are still unsure about the live video, whether the camera should have been drawn deeper, because Aldrin might be bowing down and therefore the head could be seen from above, or whether the TV camera could have been very badly levelled, then I recommend you to have a look at this video, you will find it under reference [72]. After about a minute Armstrong comes into the picture from the left and prances away to the right. Also from this movement you can see that the camera is well horizontal. Another still from the same video, this time additionally with Armstrong, is shown in Fig. 53.

Fig. 53 Another Still Image from the live TV Broadcast [72]

This observation is independent of a possible slope inclination or of an inclination of the Lunar Module. And since especially the horizon course fits very well to the image AS11-40-5928 (Fig. 43), this shows that the video and the photos were all taken at the same place. Since the video was taken in a studio, this is also true for the photos.

In summary, these photos show the landing site as a platform bordering a precipice or slope several kilometers deep to the west. This does not match the Sea of Tranquility or the Moon, but it does perfectly match a studio environment where only a limited illuminated area is seen. This area corresponds to the illuminated foreground of the reenactments (Fig. 49 and Fig. 51). The black space is actually the black wall of a studio.

Finally, let's do a short final test: in Fig. 54 I present you a picture from my vacation, where I photographed a sunrise through the dense fog. Do you notice anything? Was I able to capture a "looking down at the Sun" effect in the picture? You can find the answers in the appendix, chapter 10.5.

One can wonder why NASA only showed re-enacted images. Would the cameras have worked at all on the Moon, or would they have overheated, or would the films have suffered damage as a result of the ionizing radiation, or were the two astronauts not on the Moon at all in the end? I am trying here to provide you, as much as possible, with all the information you need to answer these questions better and better as you read. The reaction is not the same for everyone. Some forgive NASA such falsifications true to the Roman proverb "Quod licet jovi non licet bovi" which loosely translates to "What is lawful for the god Jupiter is far from lawful for an ox." So if the mighty NASA shows studio pictures in much better quality than it would have been possible with real Moon pictures, then it would still have embellished within tolerable limits. However, a mountaineer who wants to prove with a faked summit photo that he has climbed the summit would be sent packing. Decide for yourself how much disinformation you can tolerate on this subject, and whether Apollo 11 was an *Embellished success* or a *Staging*; the category of *Transparent successes* seems to me to be out of the question.

Fig. 54 Deep Sunrise

Now that we know that the pictures of the Moon landing are reenactments on Earth, we can turn relaxed to some effects that could not have occurred on the Moon or that would have turned out differently in authentic reporting. Without the knowledge of studio pictures, one would perhaps be inclined to classify some of these effects as "enhancement" or "image retouching" and to pass them over.

4.4.3 Shadows from all Sides

Probably you have heard about the discussion about parallel or not parallel shadows of the Moon images. Depending on the topography and perspective, shadows are reproduced differently. The question here is whether the shadows fit to a landscape illuminated by the Sun or not. For this I consider here *one* Moon picture representatively also for other pictures. The selected picture unites several effects in connection with shadows and lighting and is therefore very well suited for an in-depth consideration.

Apollo 11 - THE Mission

Fig. 55 AS11-40-5961 Armstrong photographs the Lunar Landing Module and his own Shadow

AS11-40-5961 (Fig. 55) shows the Lunar Module from a distance of about 60 m from the east; the astronauts did not move further away from the Lunar Module. This image combines various effects:

a) <u>Lunar background:</u> First of all: in contrast to AS11-40-5928 and to the live video here at the end of the shadow of the Lunar Landing Module further lunar background is visible. If you look closely, you can see the previous horizon line in the image. Since one looks from an elevated position - higher than at AS11-40-5928 (Fig. 48) - on the Lunar Module, one could think that one could now look over the edge of a possible hill. But since one did not see this background even in the TV picture, it was definitely not present there; there one looked down much more steeply on this horizon line. I conclude from this that one was obviously able to add this further background here photo-technically or as a backdrop. Photoshop or other electronic image processing did not exist at that time, but other methods were mastered: for example, one could project two images on top of each other, retouch them by hand, or add the background by projecting it onto a screen. Stanley Kubrick used the latter in the film "2001: A Space Odyssey" in 1968.

b) <u>Shadow of flagpole:</u> Neither flag nor flagpole cast a shadow. Flag and flagpole could have been copied into the image.
c) <u>Vanishing point in the foreground:</u> In the foreground, the shadows still converge to a vanishing point in the foreground. The vanishing point is not a sharp point, but rather a vanishing area. I have indicated the extensions of the shadows with the double-dashed lines. From the lower right, the shadow of the stereo camera comes into the picture. This camera is at an angle on the ground, but it is tilted away from the Sun quite accurately, so the shadow points in the same direction as if the camera were perpendicular to the ground. You can see this in Fig. 56, where the stereo camera is illuminated from the right; its shadow is parallel to the shadow of the surrounding stones. Its shadow points in Fig. 55 also into the vanishing area. The background with the Lunar Module is illuminated by separate spotlights and probably projected into the image; the shadows of the Lunar Module and the solar sail aim at a perspective vanishing point behind the landscape.

Fig. 56 Stereo Camera: the Shadow runs away from the Sun (AS11-40-5957)

d) <u>Lightened lunar landscape with illumination exactly from behind and halo effect:</u> These effects are on Fig. 55 more pronounced than I would have expected, but can actually occur in a landscape illuminated by the Sun. I describe this here because it gives an interesting insight into illumination conditions and is sometimes discussed in this context: a terrain illuminated exactly from behind appears brighter in the direction of illumination, if part of the light is directly reflected back, as in the case of a cat's eye or a metallic surface or if the surface has a coarse-grained structure. In the case of a coarse-grained structure, the entire visible surface is illuminated in the direction of illumination, but laterally one sees illuminated and shaded areas mixed, so that the landscape appears darker there. This is the case, for example, with the sports field at Fig. 36. There, the brightness around the shadow of the camera or around the head is at a maximum.

If the back reflection is pronounced, the head gets a gloriole. In technical jargon, such a strong brightening is called a halo. The halo effect is described by NASA „*due to some combination of diffraction around the helmet and/or the coincidence of maximum zero-phase backscatter with that part of Buzz's shadow*" ([53], at AS11-40-5882, Fig. 57). A richly complicated explanation, I think. On Earth, such an effect occurs, for example, at a crosswalk (Fig. 58). The color of the zebra stripe is mixed with micro glass spheres, which reflect the light like a cat's eye in the direction of incidence. I will come back to cat's eyes in more detail on the topic of laser distance measurement.

Fig. 57 Shadow of Buzz Aldrin with Halo Effect (AS11-40-5882)

I consider these effects (d) neutral, so neither pro nor contra studio images, because the differences in brightness are not pronounced enough.

Fig. 58 Halo Effect at a Crosswalk at Noon

4.4.4 TV Camera mounted crooked

In the chapter 4.4.2 I reported on the live video and estimated the height of the TV camera to be about the same height as the top of Aldrin's head. Here I will go into the live video again, but I will not start the observation with the video images as before, but from the position and orientation of the TV camera. Its position is best seen in the training images.

The TV camera is mounted on the MESA (Modular Equipment Storage Assembly), which is a container for equipment, as shown in the training images (Fig. 59 and Fig. 60). The folded down MESA can also be seen on Fig. 35, the only lunar image with Armstrong, standing at the MESA. The MESA is the container, which Armstrong extended as first after the exit like a folding bed, so that his first step on the Moon could be filmed by the TV camera.

Fig. 59 TV Camera on the unfolded MESA (ap11-S69-31585; the white Circle is inserted by the Author)

In the training image with Armstrong at the MESA (Fig. 60), the MESA is wrapped with insulation foil, and only the camera opening is visible. Such foil was then also used on the Apollo 11 flight to prevent the equipment from cooling down quickly after deployment. On one of the Moon images (AS11-40-5866) you can even see the wrapped camera lens when zooming in, similar to the one on Fig. 60. You can click on a link in the description of this lunar image which shows the camera lens enlarged. The section provided by NASA is shown in Fig. 61. The caption and the arrow are both inserted by NASA.

Fig. 60 Armstrong at MESA; Camera under Insulation Foil (S69-31060)

Multilayer insulation (or MLI) films are still used today to wrap parts of satellites to provide good thermal insulation.

The TV camera is at Fig. 60 about at Armstrong's chest height, which is about 1.3 m above the ground. Based on the live video, I estimated a height as the top of Aldrin's head or about 1.9 m. Let's leave that discrepancy for now. I'll come back to it shortly when we've gotten the general idea regarding the TV camera.

Fig. 61 TV Camera (a11EnhcDet5866TV labeled NASA)

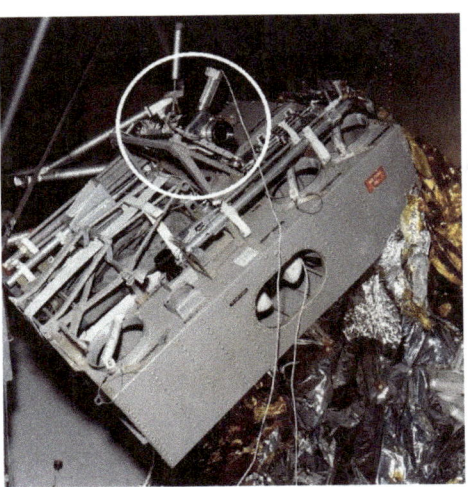

Fig. 62 The TV Camera is tilted by estimated 30° to 45° (ap11-S69-31584; the white Circle is inserted by the Author)

At Fig. 62 you can see the camera on the MESA in more detail: the top of the TV camera rests on an angled support. Accordingly, the camera handle points upward to the right by estimated 30° to 45°. In later video recordings, the camera was held in the hands or was fixed on a tripod and was operated in its normal position, i.e. upright with the handle down [AS11-40-5907, not shown here]. On the MESA, it was a better fit to fix the camera upside down. This way, the camera could be fixed in a stable position already at the start.

In the NASA documentation [53] the mounting of the camera is described as follows:„*The camera was mounted on the MESA upside down .... It was placed at an angle so that the bottom of the ladder could be viewed when the MESA was fully deployed.*" So instead of being tilted forward, the camera was mounted at a lateral angle so that the foot of the exit ladder still fit on the picture.

If the camera was operated upside down at one time, i.e., with the handle pointing upward, and at another time normally, with the handle pointing downward, the intention was to provide the viewer with an upright television image in both modes. NASA writes [73]: „*Each of the tracking stations had the capability of inverting the image so it would look normal. This was done by throwing a switch from the 'normal' position - used when the camera was on its tripod away from the LM and was, therefore, rightside up - to the 'inverted' position - used when the camera was upside down on the MESA.*" So this image inversion could rotate the TV image 180°. A normal analog TV picture starts with the first line at the upper left edge of the picture. If the picture is rotated 180°, then this first line starts at the bottom right. The electronics probably used an intermediate picture to rasterize the lines so that the picture was displayed correctly to the viewer. These electronics could rotate an image 180°, but they could not align an oblique image horizontally. The skewed assembly keeps the image skewed. You basically can't create a horizontalized image from a 30° skewed image in the same format; rotating it would add additional landscape that isn't even present in the original image, or you'd have to select a horizontalized crop.

NASA was aware that a tilted image would result even after the inversion and wrote in [74]: *"the lunar horizon is tilted down to the right by about 11 degrees because the TV camera is tilted by about that much inward toward the spacecraft, as can be seen in a detail ... from pre-flight photo S69-31585"*. (S69-31585 is Fig. 59) This sentence can only be found in versions of NASA documentation up to 2017, so I am referencing a version from the web archive (web.archive.org). This is archive software that regularly scours the web, detects changes to web pages, and saves archive copies at specific times so that you can retrieve earlier versions. So I was able to reference the November 2017 version here; after all, the current version no longer has this record. NASA may have noticed in the meantime that if the camera had been mounted at an angle when taking the picture and had therefore recorded the horizon at an angle, it would then also have had to record the astronauts at an angle.

However, as we saw in the last chapter, the video image is well horizontal, both astronauts are upright. The camera for the live video was at least at head height. In other words, the live video was taken by a well horizontalized camera (probably on a tripod) at a height of about 1.9 m, and it was not mounted on the MESA as described.

Fig. 63 summarizes what has just been described in picture form:

Left: This is roughly what the image would have looked like from a camera mounted at an angle. Both astronauts are upright. The foot of the ladder is well in the picture.

Center: Same image (only rotated) as it should have appeared on a TV.

Apollo 11 - THE Mission

Right: Official image from live video; taken from a well-horizontal camera.
I created the left and the middle image from the original on the right.

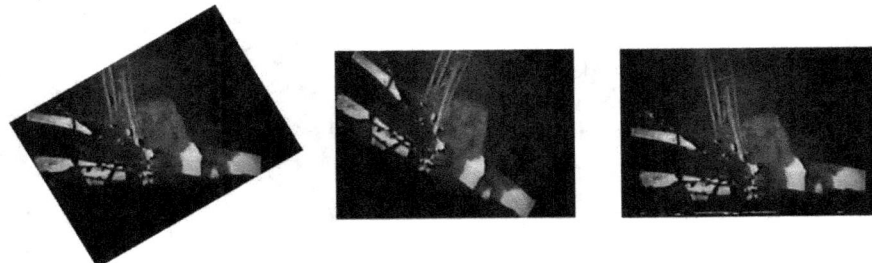

Fig. 63 TV Images with Camera mounted obliquely and horizontally respectively

4.4.5 Different Brightness of the Lunar Soil

As a result of the different spotlights aimed at the main subject, the lunar soil is illuminated differently. Even if the illumination is lateral and therefore there is no halo effect around an astronaut, even then you can see the different illumination in many pictures. This effect does not occur in a landscape illuminated by the Sun (sideways) in good weather; a flat sandy surface, for example, is equally bright everywhere.

Fig. 64 Portrait of Aldrin (AS11-40-5903)

You can see the different illumination in many images; very clear is this effect in the live video (Fig. 50), where the lunar soil on Aldrin's right is much brighter than in the background on the left. In the portrait of Aldrin (Fig. 64) the landscape in the background is much darker than the landscape on Aldrin's right. In return, the shadow of Aldrin has turned out nicely. And it is the shadow we are supposed to admire in this picture; best a second time still as a reflection in the visor of Aldrin.

4.4.6 Aldrin's Face through the Sun Visor

In Fig. 40 Aldrin is shown at the flag. If you zoom into this picture you can clearly see his face (Fig. 65). In the description of this picture from NASA it says that you can see him through the gold coated visor because he is standing at just the right angle to the Sun and tilting his head forward. I have never observed this effect with ski goggles and sunglasses on Earth.

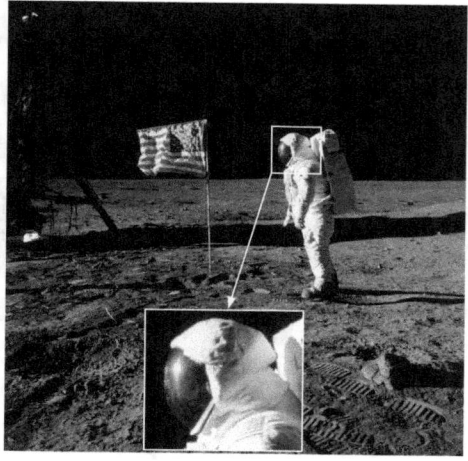

Fig. 65 AS11-40-5875 with Zoom Window (inserted by Author)

4.4.7 Life on the Moon?!

NASA has hidden various objects in the lunar images that do not belong on the Moon at all. The most conspicuous one can be seen in Fig. 66. The turtle is also present in other images of the Little West crater, for example in Fig. 56. It is interesting to observe how different people react to this image. Actually, those who believe in the *Transparent Successes* version should immediately go to the basement and get a bottle of champagne to celebrate the discovery of the first extraterrestrial life. However, I have observed a completely different reaction: for these people, this turtle is not a turtle at all but just a somewhat specially shaped rock. It is generally difficult for many people to recognize the turtle as such: in an environment where no turtle is allowed to be, there can be no turtle either.

Fig. 66 Life on the Moon? (AS11-40-5955, Zoom Window and the drawn Turtle are inserted by the Author)

Apollo 11 - THE Mission

Perhaps NASA chose a turtle because a couple of four-toed turtles had already circumnavigated the Moon in September 1968 aboard the Soviet Zond 5.

4.4.8 Press Reaction

I summarize here the reporting of two newspapers in the Zurich area, which is in Switzerland. In the next chapter, I take up some of these statements again and look at them more closely.

The Zürichsee-Zeitung dedicates the first walk on the Moon in the edition of Monday, July 21, 1969 besides the whole front page also the second page. On the front page, under the title "Walk on the Moon", there are two large pictures of the live television broadcast. Underneath, the newspaper writes that millions of radio listeners and television viewers around the world had waited in breathless suspense for the moment when Neil Armstrong cautiously opened the door of the lander and carefully descended the nine rungs to the Moon. For the first step of a man on a foreign celestial body, the control center in Houston had announced the exact time: July 21, 3:56 a.m. and 20 seconds Central European Time (CET). Armstrong had described the ground as powder-like and very soft. In the pictures of almost unbelievable sharpness the Earth inhabitants left behind could have pursued, how the 38-year-old American had done almost in the light and swinging movements of a ballet dancer his first steps on the Moon. Armstrong had reported later still that the brake engine had left small craters on the ground. Afterwards President Nixon had conveyed his congratulations in the first telephone call on the Moon. At 06.11 o'clock CET the astronauts would have closed the hatch from the inside after the re-entry and the stowage of 20 pounds in containers of packed soil samples.

On the second page, the Zürichsee-Zeitung headlines "The greatest adventure of the century has succeeded". The two subtitles read "Sunday evening, 9:18 p.m., "Eagle" landed on the Moon with amazing precision" and below that "Historic words of Armstrong: «Contact Light on, Engine Stop, The Eagle has landed»." Below that, the newspaper writes that during the critical moments of the descent, Armstrong had taken manual control of the "Eagle" to steer it away from a hilly, rock-strewn area toward which the automatic system was guiding him. One of the most important findings of the Houston control center had been the angle of inclination of the lander on the Earth's satellite: it was a little more than four degrees. At an inclination of twelve degrees, the

Apollo 11 – The Real Story

astronauts would have been lost. The ascent stage could not have ignited properly.

Even Pope Paul VI had seen the television broadcast from his summer residence Castelgandolfo and had expressed deep emotion about the successful Moon landing. The Soviet news agency TASS had briefly reported the event seven minutes after the landing of the Lunar Module and mentioned that the first astronaut was to set foot on the Moon on the morning of July 21.

The newspaper dedicates a section to each of the three astronaut women, describing their reactions ranging from excitement to tears.

"Luna 15 closer to the Moon" is the title of another segment. Luna 15 had been placed in a new orbit on Sunday; the maximum distance from the Moon was now 110 kilometers and the minimum distance was 16 kilometers, it said. The instruments on board were operating normally.

The editor-in-chief of the newspaper praised the American achievement, which 500 million people had heard about on their television sets, and described it as a victory for *men*. The whole world had a share in this flight. So the whole world should also have a share in the Moon. Who does not remember the fears darkening the early Soviet space successes that a power could try to seize our satellite star for its own purposes? Never should the Moon become a threat to the Earth.

The Neue Zürcher Zeitung (NZZ) reports in the noon edition of Monday, July 21, 1969, about the stepping on the lunar surface. The NZZ published three issues per day until 1969: morning, noon and evening. [75] The NZZ devotes the entire front page and slightly more than half of the second page to this event. The title is "The first men on the Moon" with the subtitles *"Landing of the shuttle - Armstrong and Aldrin disembark"* and "The "Eagle" touches down". Also the NZZ shows large a picture of the live TV broadcast with Armstrong's first steps, which are described as almost floating. About the landing the NZZ writes that the two lunar pilots had forgotten their usual taciturnity only for short moments: so they had learned that where the Lunar Module should be set down by the computers they had discovered a crater of the size of a football field teeming with boulders, whereupon Armstrong had taken over the manual control and selected a landing site about six kilometers further in the same approach path. The terrain on which the shuttle now stood as a ghostly foreign body in the lunar

Apollo 11 - THE Mission

landscape was practically flat with a slope of only 4.5 percent and thus fulfilled a first, perhaps decisive condition for the return to the mother ship.

Houston had announced that the *rescheduling* discussed as a possibility had become fact. The astronauts had preferred to forgo the originally planned four-hour night's rest after the thorough inspection of their vehicle and the subsequent refreshments and instead set off immediately for their "walk" on the Moon. The postponed sleep was to be made up afterwards. At 3:57 a.m. Armstrong first set his left foot on the lunar surface. Thus the climax of the mission of Apollo 11 had been reached.

Armstrong had reported after his first steps on the lunar surface that his feet sank in ever so slightly. "I can see my footprints," he had said. After inspecting the ground beneath the landed spacecraft, he said: "The recoil rocket didn't leave a crater." Shortly thereafter, he reported again, "I can see my footprints clearly in the fine sand particles. It's pretty dark here in the shade and difficult to see everything."

Further the NZZ writes that Armstrong, after also Aldrin had stepped on the Moon soil, had taken the TV camera from the mounting and had aligned this according to instructions of the ground control of Houston and had thus shown the audience the landing vehicle and sections of the lunar surface.

Under the title "Hesitant announcement in Moscow" the NZZ writes that the Moscow radio and television had managed on Sunday evening to practically ignore the *Moon landing*. No radio and television station would have interrupted its normal broadcast to pass on the news. The news agency Tass would have let the message run only 25 minutes after successful landing as a 5-line message over its ticker. Likewise in the late news service of the television first messages about metallurgy, Poland, Lenin and sport had been brought, then a film report about student demonstrations in America against the Vietnam war and only then the message about the Moon landing had been read out in the form of the Tass message, without showing even a single picture of Apollo 11.

The secret of the Luna 15 mission had still not been revealed. Only a new orbit correction was announced, after which the shortest distance to the Moon amounted to only 16 km. The fact that Luna 15 was brought so close to the Moon started again speculations about the possibility of a Moon landing. As was known, before the launch of Luna 15 Soviet sources had circulated rumors according to which the Soviets would try to get some Moon rocks before the Americans and that with the help of an unmanned spacecraft.

At 5 o'clock 05 CET (author's note: thus with about one hour delay) radio Moscow had reported the stepping of American astronauts on the Moon. No comments had been made.

4.4.9 Comment on Press Reactions and NASA Coverage

The two press articles about the Moon landing and walks are, as expected, the same in basic tenor. The Apollo program and especially the Moon landing captivated the whole Western world. The US-Americans were the heroes and their reporting was classified as extremely transparent, while one was very skeptical about the Soviet Union and accused it of dark ulterior motives such as the seizure of the Moon. The Soviet Union's coverage of this event of the century also appeared to be completely incomprehensible: according to the Zürichsee-Zeitung, a long seven minutes and according to the NZZ even 25 minutes passed until TASS reported the landing; and the television stations had not even interrupted their programs. Without saying it, it was agreed that the Soviet Union was a bad loser in the competition for the Moon.

As always with U.S. government space events, the coverage took the path from NASA over the agencies to the newspapers. There is only the one source, NASA, so we get to hear what we are supposed to hear from NASA's perspective. NASA set two priorities that many still remember today. The manual control of Neil Armstrong in the final landing phase is remembered by those who have studied the Moon landing in more detail. This action makes Armstrong a superhero: not only was he the first to set foot on an extraterrestrial object, no, he also took responsibility at the decisive moment and acted skillfully. Aldrin had to take a back seat there, too, even though he was the pilot of the Lunar Landing Module. The second focal point, the footprint, is still known today by almost the entire Western world. It became the symbol for this mission, whereby one is reminded of course at the same time, who carried out this mission. The first words of Armstrong were „That's one small step for a man, ..." and the second according to both newspapers „I can see the footprints of my boots"! In the NZZ this statement occurs twice. Armstrong then even reports that he can see his footprints perfectly well in the fine sand particles, although it is quite dark with him in the shade and therefore he cannot see everything; but the fine

Apollo 11 - THE Mission

sand particles, these he saw. The fact that it is possible to leave footprints on an alien celestial body has been emphasized by NASA ever since. Just think of the Mars rovers, where the wheel tracks can also be seen regularly on the first pictures.

To check whether some quotes were inaccurately reproduced in the newspaper articles in the heat of the moment, let's take a look at what NASA writes on its website [73]. There the communication between Armstrong, Aldrin and Houston is logged. I quote from this web page:

a) Armstrong: That's one small step for (a) man; one giant leap for mankind. (Long Pause) Yes, the surface is fine and powdery. I can kick it up loosely with my toe. It does adhere in fine layers, like powdered charcoal, to the sole and sides of my boots. I only go in a small fraction of an inch, maybe an eighth of an inch, but I can see the footprints of my boots and the treads in the fine, sandy particles.

b) McCandless (This is the speaker in the control center.): Neil, this is Houston. We're copying.

c) Armstrong: Ah ... There seems to be no difficulty in moving around - as we suspected. It's even perhaps easier than the simulations of one-sixth g that we performed in the various simulations on the ground. It's absolutely no trouble to walk around. (Pause) Okay. The descent engine did not leave a crater of any size. It has about one foot clearance on the ground. We're essentially on a very level place here. I can see some evidence of rays emanating from the descent engine, but a very insignificant amount. (Pause) Okay, Buzz, we ready to bring down the (70 mm Hasselblad) camera?

d) Armstrong and Aldrin set about lowering the camera down a cable hoist. Armstrong stands at the bottom and guides the cable.

e) Armstrong: Okay. It's quite dark here in the shadow and a little hard for me to see that I have good footing. I'll work my way over into the sunlight here without looking directly into the Sun.

So the newspapers have reported correctly. The second sentence of Armstrong after stepping on the Moon is already a preparation for the legendary footprint. NASA has prepared us for this from the very beginning. It has succeeded in placing a symbol in the Western world that everyone knows. The editor-in-chief of the Zürichsee-Zeitung still feared that the Soviet Union, if it had won the race, might seize the Moon for its own purposes. Did NASA perhaps not work with such symbols to claim the Moon for itself? I will return to this question in chapter 7.

As we have already seen with the pictures, NASA has built in contradictions again and again. The most obvious one is the turtle. This is very convenient. The normal citizen is made to believe how the astronauts walked around on the Moon by pointing out the astronauts walking around. But with the same pictures, NASA can also show people who are to be initiated where the pictures were really taken.

Armstrong's statements regarding the powdery lunar surface are another example of NASA's duplicity. As a second sentence after stepping on the Moon, Armstrong says the surface is fine and powdery. He mentions that he could see differences in the millimeter range. Shortly after, in e), he says that it is too dark for him to see if he has a good foothold. These two statements do not fit together at all: if he had really been able to judge the fineness of the powdery sand in the shade as said under a), then the rough assessment of the base would have been easy.

With the allusion to the craters the confusion game continues. The Zürichsee-Zeitung speaks of small craters under the engine, the NZZ of no craters at all, and in the NASA minutes under c) it says both, first no crater at all and then some small jets. In the description of the Lunar Landing Module [76] it says that the underside of the Lunar Landing Module is equipped in such a way that it can withstand the pressure of the return beam which is reflected from the lunar surface during landing. In addition, the underside would also be protected against the strong thermal radiation so that it could stand up to the heated lunar surface during landing. So the parts of NASA from which this description comes expected the return beam to hit the lunar surface strongly.

Aldrin said shortly before the landing at an altitude of 12 meters that they had kicked up dust. That an engine with a thrust of more than one ton blows away the sand and the dust, which Armstrong could easily whirl up, is obvious. But then Armstrong prepares us for the opposite with his statement c) that there would be no crater or only minimal jets; and on the Moon pictures there are also no traces of the engine to be seen. Thus the viewer is pointed to details, so that one does not get the idea to look at the whole. As already indicated earlier, we will have a closer look at the engine jet in the next chapter during the launch back from the Moon.

A final point is the inclination of the Lunar Landing Module. The difference between degrees and percentages in the two papers is probably due to the flurry of reporting. In the mission report [21] it says 4.5°, from which it could be inferred that one landed on an inclined slope. However, Armstrong says under c) that the landing site was level; and the word "level" means "flat" and also "horizontal". This is also a nice example of the confusion game: The Moon pictures show a level landing place. A slope or a gradient of single images in viewing direction cannot be judged in general. So a 4.5° slope in the image AS11-40-5928 (Fig. 48) might make the near horizon appear natural if the terrain rose by 4.5° towards the back and the horizon was therefore higher than the photographer's position. The other photos and the statement of Armstrong would have to be omitted then, however, obligingly. In such a way most of us were fooled for a long time or are still fooled today.

4.5 Launch back from the Moon and Rendezvous with the Orbiter (CSM)

On Monday, July 21, 1969, at 6:11 a.m. CET, the astronauts have completed their external activities by closing the hatch, then catch up on their rest and lift off from the Moon on the same day with the ascent stage alone at 6:54 p.m. CET. The descent stage serves only as a launch pad and remains on the lunar surface. At the time of launch, the Command and Service Module CSM will be in its near-circular orbit of 60 miles or 111 km altitude, in exactly the right position for the intended rendezvous. [21] Fig. 67 shows the trajectory of the ascent stage from point 1 to 9 to the orbit of the CSM. Docking is then completed at point 10.

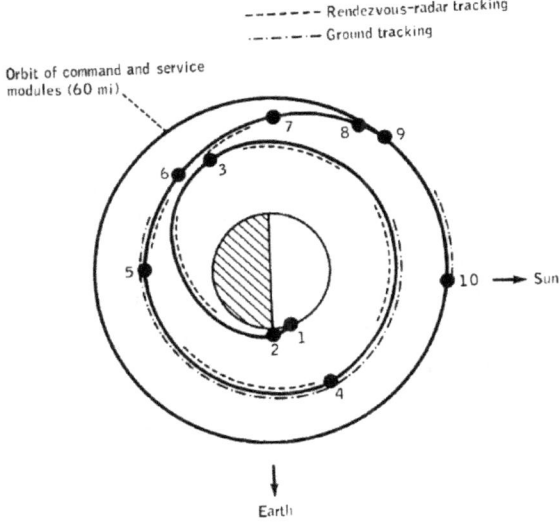

Fig. 67 Ascent and Rendezvous Trajectories [21]

The thruster of the ascent stage can neither be swiveled nor throttled [77]. This means that it produces a constant engine jet to the rear and thrust in the opposite direction. The engine jet does not need to be throttled during ascent, since one is always at full throttle. During takeoff, the engine jet is directed downward and the thrust is therefore upward. The engine burns only in the very first stage, i.e. from takeoff (point 1) to point 2 during 7.15 minutes up to an altitude of 18 km and a speed of 1.7 km/s. [78] The ascent stage is then on an elliptical orbit that takes it to point 3 to an altitude of 83 km. [77] As the motor was rigidly mounted, it could not be used to change direction as in all previous stages. For control, the ascent stage used attitude control nozzles, of which it has four groups of four nozzles each (Fig. 18). This works here because the ascent stage is relatively light and the attitude control nozzles are now generously dimensioned because they had to control the entire Lunar Landing Module, i.e. ascent and descent stage together, during the landing. Since the two astronauts are not strapped in and are standing, the attitude control jets must also compensate for any weight shifts of the astronauts. From point 2, the astronauts are mostly weightless. Only at points 3 to 9 are short course corrections and acceleration maneuvers made, with maximum velocity changes per maneuver of 16 m/s [21]. I have already described in the chapter 4.2 that in the lower lunar orbits the velocities are similar in each case, so that here only small accelerations are necessary. So the attitude control system is sufficient for this as described.

From launch to docking takes 3 hours 41 minutes; the CSM has made slightly less than two orbits of the Moon in that time. After docking, Armstrong and Aldrin join Collins in the CSM and also stow the lunar rocks and solar sail in the CSM. Afterward, the ascent stage is no longer needed and is detached. The ascent stage will remain in this orbit [79] and should still be circling around the Moon if it was not deflected by a heavy meteorite. Alternatively, the ascent stage could have been allowed to crash into the Moon. For this it would have had to reduce its speed by only 25 m/s. With a 25 m/s reduction, the ascent stage would then have crashed tangentially onto the lunar surface; to achieve a steeper impact, the braking could have been applied a little more. Compared to the other maneuvers, the additional fuel consumption would have been acceptable, and with Apollo 5, NASA had demonstrated that the ascent stage could also be operated unmanned. Today, satellites that have completed their mission are brought down on the celestial body under investigation, thus keeping the orbit free of space debris and not endangering future missions. For example, NASA let LADEE, a satellite that had been studying the Moon's atmosphere, crash into the far side of the Moon in April 2014 after a 7-month lifespan [80] [81]. The alternative, flying off the Moon altogether, would require much more energy. The ascent stage would have had to increase its velocity from 1.6 km/s to the escape velocity of 2.3 km/s from its orbit at an altitude of 111 km to move away from the Moon forever. The attitude control system would not have been sufficient for this.

The described flight path from launch to rendezvous fits the physics of space flights. Nevertheless, it is worthwhile to take another closer look here and to compare this phase with today's state of the art. For now, we will look at the flight path and the rendezvous maneuver; the actual launch phase, i.e. the first 10 seconds, I will deal with at the end of this chapter.

Although the trajectory from launch on the Moon to the CSM, which waits in a roughly circular orbit of 60 miles or 111 km, could be imagined to be even shorter, namely if the ascent stage had accelerated more at point 3 and flown directly up to the CSM orbit. Its flight would then have taken it below points 7, 8, and 9 on the shortest path to the CSM orbit. However, an additional loop was made. The direct ascent would only have made sense if the CSM had been at the right place and if everything else had fit. However, the CSM was ahead and the ascent stage had to catch up with the CSM first and then accelerate at the right moment and fly up to the CSM. During this extra loop, the ascent stage had time to measure its own trajectory and the direction and distance to the CSM so that it would actually meet the CSM on the higher trajectory. To do this, the astronauts used a sextant, a distance measurement system and the rendezvous radar.[21] To make a successful rendezvous in the end, the position and velocity of the two spacecraft must be known very precisely, and the approaching body requires fine and precise control of its propulsion thrusters. We have already seen with the orbital velocity that with the CSM orbit a velocity too low by 25 m/s leads to the fact that the spacecraft oscillates back and forth on an ellipse between 111 km altitude and the lunar surface. With an absolute velocity of 1.6 km/s, 25 m/s are just 1.6%, and the influence of this 1.6% is huge as seen. So you have to have the velocity very well under control and also be able to control it precisely in small steps, not only the amount of the velocity in meters/second but also the direction.

These high demands on the control systems of the spacecraft and the measurement of the orbit data have led in the time after Apollo in space travel to the fact that one needed from the start to a rendezvous in the Earth orbit two to three days, whereby one can use radar measurement from the ground on Earth. Radar surveying from the Earth was also used on the Moon, but due to the great distance it was less accurate and only possible at all on the front side of the Moon.

Usually, during a rendezvous, one spacecraft plays the "hunter", that is, it actively maneuvers to the "hunted", which passively flies on its orbit. Here, the ascent stage was the hunter chasing the CSM, with the CSM cooperating and adjusting its trajectory several times. As the most prominent example above the Earth, the ISS is chased, today by Soyuz spacecraft and earlier by the Space Shuttle. The Space Shuttle was also operated by NASA, which should know all the tricks for a quick rendezvous. Surprisingly, even in 2010, the Space Shuttle did not take three to four hours from launch to rendezvous like the ascent stage at the Moon, but about three days. [82] The Space Shuttle had to circle the Earth many times, adjusting its trajectory in small increments. Until 2013, it would always take about 50 hours, or a little more than two days, to get to the ISS. On March 29, 2013, this time was significantly undercut for the first time with just under 6 hours. On the one hand, the launch took place at exactly the same time as the ISS passed over the Baikonur launch site, and on the other hand, the propulsion and control systems had been improved. [83] Among other things, the Soviet/Russian *Kurs* docking system was used, which uses several radar antennas to determine the relative position, attitude and rate of approach and then steers the spacecraft accordingly. [84]

In my estimation, it should take about the same amount of time from launch to rendezvous on the Moon and on Earth. The radius of the Moon is only about 1/3 of that of the Earth and the gravitational pull is smaller, so on Earth stronger thrusters are needed to get into a first orbit. However, this took Saturn V only about 12 minutes. After that, the ratios are close: the orbital periods are similar at 90 and 120 minutes, and so is the maneuvering on these orbits. One also needs about the same velocity corrections to raise the altitude of an orbit for example by 10 km, because the differences of the orbital velocities are almost the same: at the Moon the orbital velocity of a 110 km orbit is 4.4 m/s smaller than of a 100 km orbit and at the Earth this difference is with a

Apollo 11 - THE Mission

typical orbital altitude of 300 km with 5.8 m/s similarly large. So we are moving in the same orders of magnitude and as just said, the time to the first orbit is very short with 12 minutes. That's why the stated 3 hours 41 minutes for Apollo 11 for the return launch from the Moon to docking surprised me a lot, especially when I compared this number with the 50 hours on Earth, which was standard until 2013. Did I miss something that makes it so much easier on the Moon? Especially the orbit corrections must have been rather inaccurate on the Moon, they had much less sensors and the fineness of the control was also still on a simpler level. The manual control by the astronauts could have brought advantages only in the last phase with the docking; with the course corrections also the astronauts were dependent on radar measurements and other sensor data.

Another comparative flight was the Apollo-Soyuz project, when in July 1975 an Apollo and a Soyuz capsule demonstrated a rendezvous in an Earth orbit of 229 km altitude. The Apollo capsule, which launched second, took about 44 hours to dock, with both capsules playing "hunter" and repeatedly adjusting their orbits for docking. [85] So again, NASA was nowhere near being able to reproduce the 3.41 hours from the Moon. As I remember, they were very proud at that time that the docking worked out at all and that the two astronauts were able to perform the "most expensive handshake" so far.

Why does it still take five to six hours to get to the ISS these days in the event that the launch time is set as conveniently as possible? The European Space Agency ESA has produced a video that explains the rendezvous problem very nicely [86]. A Soyuz capsule is initially placed in an initial orbit of 220 km altitude, while the ISS typically flies at 400 km altitude. The transition from the initial orbit to the ISS orbit cannot be done in one step today. First, one flies to an intermediate orbit (phasing orbit) with a precisely defined altitude and only then ascends to the ISS orbit during 1 ½ Earth orbits. This is a far cry from the very direct approach of Apollo 11.

NASA practiced and tested the rendezvous maneuvers on the previous Apollo flights and also on the missions before that. I have already mentioned coming back to the rendezvous maneuver on Gemini 8 and Gemini 11. For Gemini 8, it took 6 ½ hours from launch to docking, and the initial constellation was not optimal and Gemini 8 had to correct its orbital plane. Today, it continues to take 50 hours to rendezvous under these conditions. Gemini 8, by the way, was the first space flight by Neil Armstrong and his partner David Scott. Gemini 11 was even better: Gemini 11 took an incredible 94 minutes from launch to docking, which means the target satellite made just one single orbit around the Earth in that time. It was launched 90 minutes or one orbit before Gemini 11, as was the case with Gemini 8.

Before we take a closer look at such a fast and direct rendezvous of spacecraft orbiting the Earth, I will demonstrate this with two aircraft: if, for example, a suspicious transport aircraft is now flying south at an altitude of 10,000 m over Zurich, and Switzerland wants to control this transporter with an interceptor, then this can be done most quickly if the interceptor also takes off and climbs south at Zurich right now. This way, it meets the transport aircraft at an altitude of 10,000 m and has directly the same speed, and both fly south. So the interceptor can now calmly circle around the transport aircraft and inspect it. If you transfer the flight tracks to a map, it is the same line going south from Zurich for both cases. If the two airplanes could fly straight on and on, then after a long time they would fly over the South Pole, later over the North Pole and then come back over Zurich from the North.

It would be different if the interceptor started 100 km west of Zurich in the Jura. Then it would fly to the east or slightly southeast. It would have a longer route and would have to make at least one turn in order to fly south at the same altitude as the transport aircraft at the end and to be able to inspect the aircraft over a longer period of time.

Many things are similar for spacecraft. For a good overview, one likes to display flight tracks of satellites like those of airplanes on a world map. Likewise, the convergence is similar: it is also fastest in spaceflight to climb up to a satellite when it is flying over you, i.e. when the flight tracks coincide on the map. If you then start in the same direction, you only need to climb up and ideally you will meet the satellite at the top. If we consider that the launching spacecraft needs some time to reach its speed, then it is better if the overflying satellite has not yet reached the launch site completely, i.e. is a bit behind. Nevertheless, the two flight tracks coincide on a world map. Fig. 68 shows a side view of how the two satellites find each other when one flies over the launch site and both fly in the same orbital plane and therefore have the same flight track on the ground. The flight

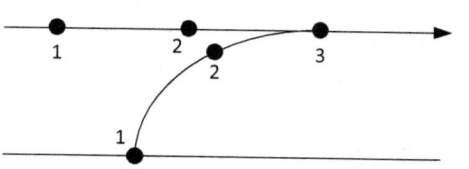

Fig. 68 Ascent to a Satellite flying over the Launch Site (1 below)

Apollo 11 - THE Mission

track is the lower horizontal line. At takeoff, both spacecraft are at their point 1, a little later at points 2, and then together at point 3.

If the flight path of the satellite, which is already in orbit, passes *aside from* the launch site, then the "chaser" must make a lateral correction in order to come onto the flight path of the "chased" satellite. Lateral corrections are tried to be avoided in spaceflight, because they usually require a lot of fuel, as we will see in a moment with an example. The maneuvers we have looked at so far for the lunar orbits were all in the same orbital plane and required very little propellant. The orbits could be easily plotted on a sheet of paper; the drawing plane and the orbital plane were identical. We did not need perspective views before.

In the previous example with the transport aircraft over Zurich and the interceptor with takeoff in the Jura, the interceptor had to fly at least one turn, i.e. make a lateral corrective movement, in order to be able to fly nicely alongside the transporter. This curve would be smallest if the interceptor took off exactly to the south and met the transport aircraft only at the South Pole. Then the interceptor would only have to fly a very small right turn and could then fly right along with the transporter - to the North Pole and back to Zurich, always along the longitude of Zurich. For airplanes, this example seems strange, because they can't fly that far in one piece and because flying straight for a long time requires a lot of aviation fuel. But I chose the example with respect to spacecraft. Once spacecraft are on their orbit, they continue in free fall on their orbit without consuming fuel. If they were both on a course from Zurich and from the Jura respectively, then they would only have to wait half an hour and they would meet at the South Pole and the hunter could change to the flight path of the hunted with a small course correction.

With now a better idea of satellites flying side by side, we return to Gemini 11 and the Agena rocket launched 90 minutes earlier. One might think that after one orbit the Agena rocket would just fly over Cape Canaveral again and that therefore the ideal constellation for a quick rendezvous would be present. But this is not so, as we have already seen with the orbit of Gagarin. If an airplane flies from Zurich exactly to the south and then via North Pole back, then it arrives after an Earth orbit again with Zurich. The atmosphere rotates with the Earth and therefore we do not perceive the Earth's rotation, neither on the ground nor in an airplane. But if a satellite flies over Zurich to the South Pole and then over the North Pole back to the South, then the Earth has rotated to the East at a speed of 15° per hour after one orbit, i.e. 22.5° in the 90 minutes. Since Zurich is located at 8.5° east longitude, the satellite would pass Zurich after one orbit at 14° west longitude, i.e. in the Atlantic Ocean west of France. The lateral distance to Zurich would be 1,700 km. With Gemini 11 and the Agena satellite it was similar, as Fig. 69 shows. Trajectory 1 is that of the Agena rocket, which was launched first. Trajectory 2 belongs to the Gemini 11 space capsule. Trajectory 2 is 22.5° east of trajectory 1, because the Earth also rotated 22.5° between the two launches, shifting the launch point 22.5° to the east. Fig. 69 shows the two trajectories with their respective starting point in Cape Canaveral. You can see that from any point of trajectory 1, if you go 22.5° to the east, you get to trajectory 2; this is true even for the intersection of the two trajectories. Now, if the Agena rocket were exactly one orbit ahead, it would be 22.5° west at the launch of Gemini 11, so it would be just back at its launch point. In reality, however, it would have to be a little more than one orbit ahead so that it would hit Gemini 11 at the intersection of the trajectories, as we will see in a moment.

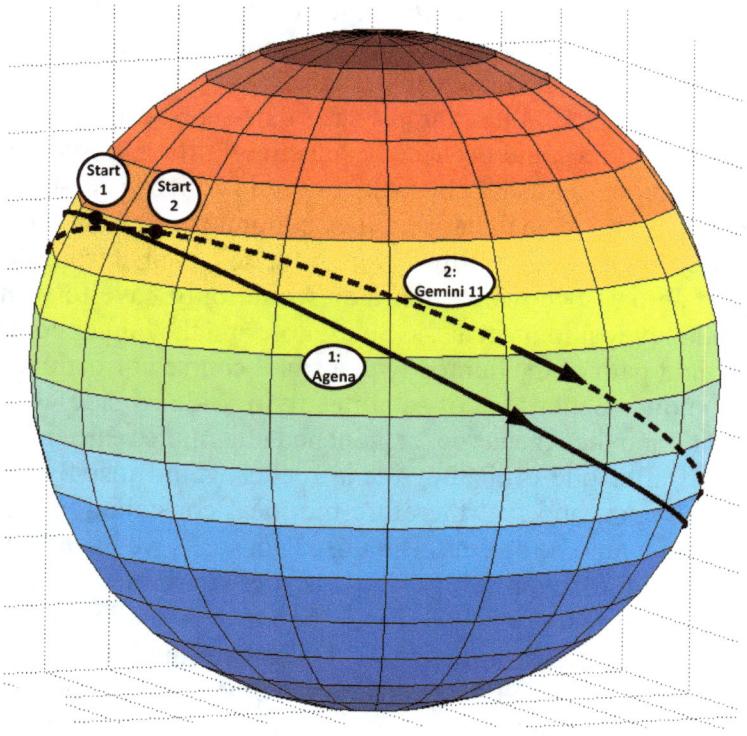

Fig. 69 Two Orbits offset by 22.5°

Both trajectories have an inclination of 29° with respect to the equator. Thus, they cross the equator in Fig. 69 at an angle of 29° and

Apollo 11 - THE Mission

reach a maximum latitude also of 29°. This inclination matches the launch site at Cape Canaveral, which is at 29° north latitude. This inclination results when the rocket launches to the east and can thus take as much momentum as possible from the Earth's rotation. A point on the Earth's surface at 29°N has a velocity due to the Earth's rotation of about 400 m/s.

Now, if Gemini 11 is flying in orbit 2 after launch and wants to make a rendezvous with the Agena target satellite, which is on trajectory 1, Gemini 11 must change its flight direction. Gemini 11 does this best where the two trajectories intersect, as shown in Fig. 70.

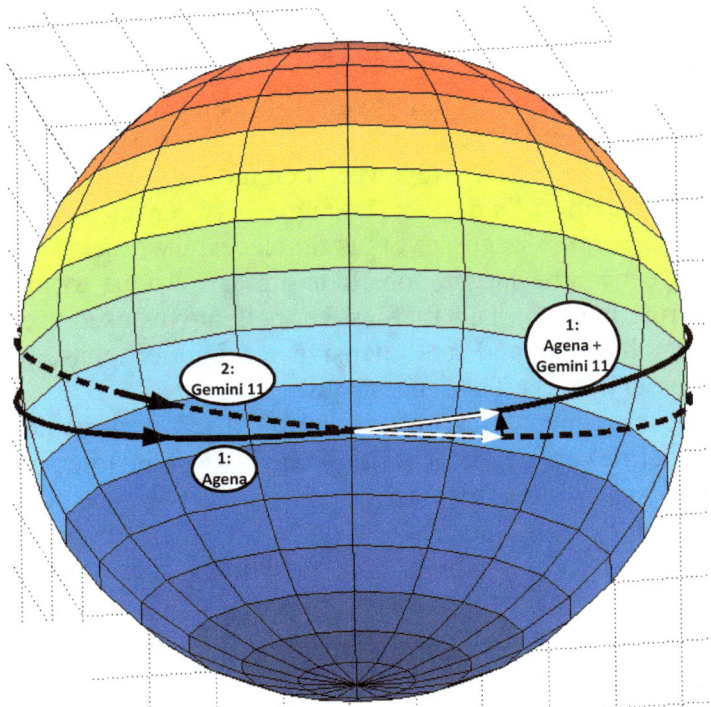

Fig. 70 Intersection of the two Trajectories

I have shown in Fig. 70 at the intersection point the velocities of the two spacecraft with white arrows or vectors. If Gemini 11 now wants to correct its trajectory around the intersection point to trajectory 1, it must change its velocity around the short black arrow upwards respectively northwards, i.e. Gemini 11 must thrust or accelerate in the direction of the black arrow. The angle between the two velocities is 11° and the length of the black arrow is 19% of the length of the velocity arrows. This means that Gemini 11 has to correct its velocity of about 7.8 km/s by 19% or 1.5 km/s to get to orbit 1. Its velocity will then still be 7.8 km/s, but pointing in a different direction, so that Gemini will then also continue on orbit 1. This example shows that changing orbital planes can be very fuel-intensive, unlike the maneuvers described so far, which involved adjusting orbital altitudes. Since such a large maneuver can only be executed imprecisely, it requires subsequent course corrections to ensure that Gemini 11 flies exactly on orbit 1 and level with the Agena target satellite. As we see right here, sometimes you have to wait half an orbit until the constellation is such that you can make the next correction. And to determine the exact correction, one should measure over a longer time and then correct in small steps. The whole thing takes time. There is still no practicable solution for a fast rendezvous today.

Based on Fig. 69 alone, one might think that Gemini 11 could have started a little more to the south instead of to the east, to get from the start to trajectory 1 faster. However, then its trajectory would have had a greater inclination than 29°. The angle between the two trajectories would have become larger and thus the necessary trajectory correction would have become larger, too. It is almost curious to mention that NASA wrote in Gemini 11 that less propellant was used than expected [13] although a long and fuel intensive acceleration phase would have been necessary for the fast rendezvous.

For the subsequent docking, as mentioned above, it would be ideal if the two satellites arrived almost simultaneously at the intersection of the two trajectories, so that after the velocity correction, one could proceed directly to docking - theoretically, at least. Such a coordination was roughly done by NASA by choosing two launches staggered by 90 minutes and one orbit, respectively. Since NASA didn't mention at all that the earth is spinning away, it was also fitting that the launches were offset by exactly one orbit. NASA could sell this to the audience by not even bringing up that the Agena rocket had not yet reached Cape Canaveral again at the launch of Gemini 11 and would then pass southwest of Cape Canaveral a little later. So the reporting fits an Earth that does not rotate and a perfect ascent trajectory as this could never be achieved until today - except for Gemini 11.

Let's summarize again: Since 2013, the best time from launch of a manned spacecraft to docking in an ideal initial constellation, i.e., when the target satellite is just passing over the launch site, is five to six hours (as of 2018). In a general initial constellation, as was the case for Gemini

Apollo 11 - THE Mission

8 and 11, it still takes 50 hours, over 30 times longer than the 94 minutes reported for Gemini 11.

NASA mentioned a completely unrealistic time from launch to docking when reporting on Gemini 11. Since the Gemini 11 mission lasted barely three days, one would have been lucky to make a rendezvous at all in that time. Therefore, there had been no time at all for all the feats that were otherwise reported on Gemini 11. When I discovered the Gemini 11 website while writing this book, I was very surprised to see what fantastic numbers and stories NASA had published here. Either NASA had hoped to be able to make such fast rendezvous in the future, or their text writers simply wanted to emphasize how good NASA was at spaceflight, including the maneuvers involved. It crossed my mind that if I mentioned all this in my book, the critical reader would surely check this reference as one of the first. Now if NASA had deleted or changed this web page in the meantime, I would be accused of improper reporting. So I immediately took a screenshot. Lo and behold, NASA did indeed delete that page during 2018. But I still found it in the web archive and adjusted the reference accordingly [13]; I am curious if it will disappear there as well. Covering all traces might be difficult even for NASA, since these maneuvers are also described in Wikipedia, for example. [87] After all, NASA seems to have realized that such information makes them completely untrustworthy.

But back to the reporting of Apollo 11. Also during the return flight from the Moon, NASA did not simply present the time from launch to rendezvous a bit too optimistically. If the astronauts had needed more than the reported 3.41 hours, they would have landed on Earth correspondingly later.

Gemini 8 was the first time ever that two spacecraft docked with each other in space. The reported time from launch to docking of 6 ½ hours is eight times shorter than it takes today. The Agena rocket had been launched 1.40 hours before, so that also here a general initial constellation was present, with which one needs today still well two days.

In the appendix (chapter 10.7) I still examined whether the Agena rocket could have launched into a different orbit than reported, so that Gemini 11 could have ascended into the same orbit. Indeed, this possibility exists, but it would contradict the documentation and would hardly change the matter, since even for the ideal initial constellation, where the target satellite just overflies the launch site, it took at least two days to dock with the target satellite until 2013.

Apollo 11 – The Real Story

The conclusion from this now seems trivial to me: NASA simply lied about the times from the launch to the rendezvous of Gemini 8 and 11 and the return launch of Apollo 11 from the Moon and told us made-up stories.

At the end of this chapter we return as promised again to the launch phase, i.e. the return launch from the Moon. In contrast to the too short rendezvous maneuver, this launch phase is more generally known and also gives rise to discussions again and again. It consists of a vertical ascent phase to get away from the ground and the launch pad. Thus, within the first 10 s, the ascent stage reaches an altitude of 75 m and a speed of 15 m/s. [28] This corresponds to a constant acceleration of 1.5 m/s^2 or 15% of the acceleration due to Earth gravity. Together with the lunar gravity of 1.6 m/s^2, the two astronauts feel a total acceleration of 3.1 m/s^2, or about 1/3 of the acceleration due to gravity on Earth. After the first 10 seconds, the ascent stage slowly tilts to the west to enter the orbital plane of the CSM.

The ascent stage is the upper part of the Lunar Landing Module; it sits on top of the descent stage, which, as its name implies, was only used for descent. The descent stage is now only a launch platform for the ascent stage. Shortly before the relaunch, the ascent and descent stages are separated from each other: the mechanical connections are released by detonating the explosive bolts placed there, and in addition, all electrical and hydraulic lines between the ascent and descent stages are cut. Then, powered by its rocket motor, the ascent stage can lift off and make its way to the CSM as described above. [76] I would like to clarify two things that are sometimes heard. First, the ascent stage ascends only by virtue of its rocket motor. There is no spring assembly between the two stages that would catapult the ascent stage up the first few feet. And secondly: the descent stage remains passive. In particular, its engine also remains where it is; it is not jettisoned to create a clear channel for the return jet of the ascent stage.

Now, when the upper stage, sitting on the lower stage, ignites its rocket motor after cutting all the connections, the upper stage lifts off, but the hot gases of the back jet cannot escape freely. The description of the Lunar Landing Module [76] states that the return jet of the ascent stage is deflected by the lower stage in such a way that it can escape between the two stages. How well the bottom of the ascent stage could be protected from the reflected exhaust gases is not mentioned in this description. Ironically, it only says that the *descent stage* was well

Apollo 11 - THE Mission

protected from the reflected gases. But this could have been burned down, since it was no longer needed. As mentioned in the chapter 4.4.9 the description of the Lunar Module says that the bottom of the Lunar Module (i.e. the bottom of the descent stage) was equipped in such a way that it could withstand the reflected back jet from the lunar soil; and in this case the back jet had much more space to escape than between the two stages.

It seems to me that there is a great danger that during the return launch from the Moon the bottom of the ascent stage will become too hot and at most partially melt. I am not aware of any comparable rocket launches, also no test on the ground. Of the 16 according to Wikipedia built Lunar Landing Vehicles apparently never one was diverted for an upper stage launch test. The hot gases from the return jet need a clear channel to escape and therefore not destroy the bottom of the spacecraft. Even missiles fired from submarine silos do not ignite their rocket motor inside the silo, which is closed at the bottom. Although the silo would act like the barrel of a cannon at first, and the resulting gas pressure would further accelerate the missile, the heat and pressure could destroy the missile from below. The missiles fired from submarines are first catapulted out of the silos by means of gas pressure and only then ignited. [88]

So the ascent stage sat on a launch platform as it had never been used before and where the exhaust jet could not escape downward. With Apollo 11, this was not particularly thematized at that time. According to NASA reports, the launch went according to plan, and access to the detailed descriptions was not easy at that time without the Internet. During the Apollo missions 15 to 17 NASA showed the launch of the ascent stage by TV transmission. With a color TV camera left behind, the launch was filmed remotely. We look at this on the example of Apollo 17 with the help of the video of the Lunar Surface Journal [89]. The first phase of the launch looks like an explosion caused by the ignition of the engine: the invisible engine jet ejects insulation material of the descent stage sideways between the two stages. Then the upper stage floats away as if guided by magic.

The following figures show the launch of Apollo 17. The ascent stages were essentially identical in construction, so Apollo 17 lifts off like Apollo 11. The video begins 9.2 seconds before the engine ignites. Nothing happens until 9.1 seconds. At 9.2 seconds, it glows red between the ascent and descent stages; I take this as the time for ignition. Fig. 71 until Fig. 73 show the start sequence: 0.3 s after ignition you see the isolation of the descent stage flying away; I take the pictures after one and two seconds for estimating the acceleration.

Fig. 71 Re-launch of Apollo 17 (0.3 s after visible Ignition)

Fig. 72 Re-launch of Apollo 17 (1 s after visible Ignition)

Fig. 73 Re-launch of Apollo 17 (2 s after visible Ignition)

According to the Press Kit [28] the ascent from the start takes place with a constant acceleration of 1.5 m/s², so that the ascent stage has risen by 0.75 m after 1 s and by 3 m after two seconds. In the mission report [21] is a graph which confirms this and also shows that the Eagle here flies up evenly with constant acceleration. In the video, however, the ascent stage flies away much faster, having climbed an estimated four meters after one second and about 10 meters after two seconds. This corresponds to a much higher initial acceleration, which then decreases again. One could now try to explain this by an "air cushion effect" of the reflected engine jet, which additionally catapulted the ascent stage upwards. The acceleration necessary to ascend to the four meters with "air cushion effect" in one second is 80% of the acceleration due to Earth gravity, i.e. 0.8 g. This would have given an unpleasant jerk, especially when the two astronauts were standing upright in the cockpit. According to their conversation, however, they experienced a smooth launch, with Aldrin saying two seconds after liftoff: „We're off. (Pause) Look at that stuff (insulation from the decent stage) go all over the place. Look at that (LM) shadow. Beautiful." [90]

Apollo 11 - THE Mission

Another anomaly is the missing or invisible retro-jet. I have addressed this with the emblem of Apollo 10 (Fig. 10), where a nice return beam is drawn at the ascent stage. It is the most natural thing in the world that hot gases are hot even in vacuum and therefore glow. The Sun and its corona shine too. But if you put yourself in the position of NASA at the end of the 1960s, you can understand why the back ray was omitted. In the studio, it would have been costly to produce a nice return beam, and besides, one would have run the risk of scorching the descent stage if you handled fire. And as mentioned earlier, Photoshop or other image editing programs didn't exist back then. Photography and TV worked analog and digital images didn't exist yet either. So they left out the reflection and explained to us that it was invisible. This explanation also fitted perfectly to the missing traces of the return jet on the lunar surface after the landing. By the way, 11 seconds after launch, when the ascent stage tilts towards the west and you can see the stage from below, you can see a bright filled circle. So the "exhaust gases" are directly visible from behind respectively here it was film-technically well feasible to add a bright circle or to produce the effect otherwise. The Lunar Module (engine) plume is mentioned in Apollo 17 shortly after the video in the text of the surface journal. [89] It is said to have interrupted the reception signal from the ground station.

After we were presented with only studio shots of the Moon on Apollo 11, this was no different on Apollo 12 to 17. I do not go into it further in this book, since many parts of the coverage are basically the same, so it would be boring to look at the following missions in the same detail. Only the return launch of Apollo 17 provides two new confirming elements for a studio environment: the too fast ascent in the first seconds compared to the written documentation and the missing engine flame.

In summary, the return launch from the Moon and the subsequent rendezvous maneuver seem to me to be very far away from the reality of the space technology of that time, which would have been needed for a Moon flight with subsequent return. This is most impressively confirmed by the time from launch to rendezvous of Gemini 11, which is 30 times shorter than what can be achieved today. And if already with Gemini 11 the rendezvous was only faked, so the conclusion is obvious that the reported super time of 3.41 hours at the return launch from the Moon is also fictitious.

4.6 Return Flight to Earth and Splashdown

On Monday evening, July 21, 1969, the Lunar Module docked with the CSM and the two Moonwalkers transferred. The three astronauts are now all together again in the Command and Service Module CSM, neatly stowing everything brought from the Moon in the Command Module. Two hours after docking, the Lunar Module is detached and another 20 minutes later, the CSM flies away from the ascent stage to have a clear path for the next and final maneuver in lunar orbit. The astronauts perform this maneuver after 2 ½ orbits of the Moon on Tuesday, July 22, 1969 at 5 a.m. Universal Time, which means the Service Module engine fires for a final time during 2 ½ minutes behind the Moon, accelerating the CSM toward Earth. [21] The last 60 hours of the mission have begun.

For the time being, the journey now continues under weightlessness in the direction of Earth. Only one course correction is necessary in total. Half an hour before landing, the Command and the Service Module separate, and the Command Module alone prepares for a controlled entry into the Earth's atmosphere. The Service Module is no longer needed and burns up in the Earth's atmosphere. Fig. 74 shows this sequence schematically: at the top, the CSM fires to accelerate toward Earth, and at the bottom, it is only the Command Module that sets off for a controlled landing.

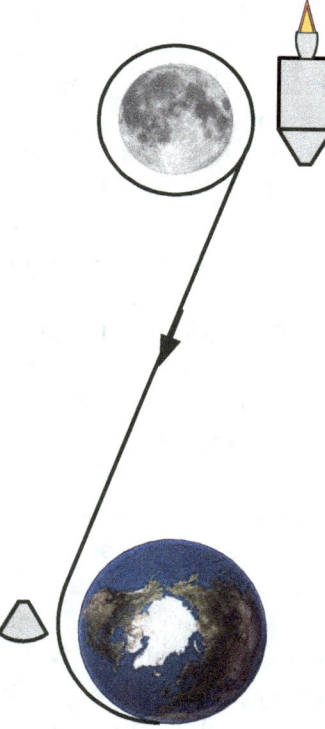

Fig. 74 Trajectory and Configuration when Leaving Lunar Orbit and shortly before Landing

The astronauts, lying in the Command Module, enter the Earth's atmosphere with a maximum speed of 11 km/s. As expected, the Command Module is catapulted upwards again slightly twice, but without being thrown away from the Earth again. According to the mission report [21] the maximum deceleration is 6.5 times the acceleration due to Earth gravity. Thanks to its own attitude control system, the Command Module flies nicely through the atmosphere with the aft heat shield (the larger one) in front. Fig. 75 shows the entry trajectory; I added the 50 km altitude line.

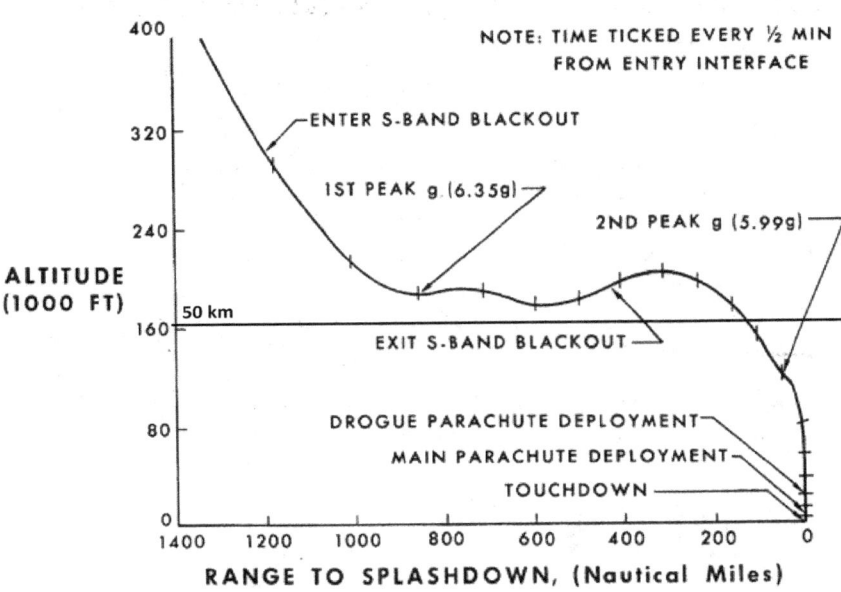

Fig. 75 Entry Trajectory of the Command Module [28]

At an altitude of 24,000 feet or 7,300 m, the forward heat shield (the smaller one) is jettisoned. Subsequently, two drogue parachutes with a diameter of 5 m are ejected. This will further decelerate the Command Module and stabilize it in its flight attitude. The drogue parachutes are jettisoned shortly afterwards and the three main parachutes with a diameter of 25 m are ejected. The impact velocity on the water with the three parachutes is 9.4 m/s. This corresponds to a free fall height of 4.5 m. Two parachutes would have been sufficient for a safe landing. [28]

The landing takes place shortly after daybreak on Thursday, July 24, 1969 at 16:51 Universal Time in the middle of the Pacific Ocean: 13°N, 169°W.[21] Since takeoff, 8 days and a good 3 hours have elapsed.

The Navy sends an inflatable boat to the Command Module after landing, the three astronauts climb out and board the inflatable boat independently. Fig. 76 shows them in the inflatable boat together with a Navy diver. You can also watch the exit on a YouTube video. [91] The eight-day weightlessness is not noticeable to them. They are already wearing a protective suit to prevent contamination with biological germs. In the inflatable boat, they are then cleaned with an iodine solution.[92]

Fig. 76 The three Astronauts in the Rubber Boat (ap11-S69-21698, Detail)

Fig. 77 The three Astronauts between Helicopter and Quarantine Container (ap11-S69-40753)

From the inflatable boat, the three astronauts are flown by helicopter to the aircraft carrier U.S.S. Hornet. There they go into a quarantine container where they are subsequently taken to a laboratory in Houston, Texas. Fig. 77 shows the three astronauts, how they go on U.S.S. Hornet straight from the helicopter to the quarantine container. In this container they are greeted by President Richard Nixon on the same day (Fig. 78).

The Command Module was taken aboard the aircraft carrier about three hours after the splashdown. [21]

Also in this last phase of the mission, two points seem to me particularly noteworthy. The speed at which the Command Module enters the Earth's atmosphere, 11 km/s, is significantly higher than spacecraft traveling from Earth orbit at somewhat less than 8 km/s. Space Shuttles sometimes lost tiles, and the Space Shuttle Columbia broke apart on re-entry into the atmosphere on

Apollo 11 - THE Mission

February 1, 2003. That things always worked out so well for the Command Module seems miraculous by comparison.

But what surprised me the most was that the three astronauts left a very cheerful impression already when they left the Command Module, then when they got into the rubber boat and finally when they left the helicopter. The eight-day weightlessness was not noticeable at all. By the way, this was standard in the sixties, already the crew of Apollo 7 had received congratulations on the recovery ship less than an hour after landing (Fig. 9).

Fig. 78 Armstrong, Collins and Aldrin (l.t.r.) and U.S. President Richard Nixon (ap11-S69-21365)

After we have seen in the meantime that all photos of the Moon originate from a studio and the three astronauts would not have had to fly also to the Moon, in order to be able to present all the pieces of evidence, there one could ask oneself whether they simply circled around the Earth during 8 days. But also then they would have experienced 8 days of weightlessness and would have climbed hardly so lively from the Command Module. Therefore one could make the assumption that the three astronauts had remained on the ground during the launch of the Saturn V rocket and the rocket had taken off unmanned. For the splashdown, the Command Module would then have been tipped out of an aircraft and could have landed or splashed down on its three parachutes. In this way, NASA would have reduced the accident risk for the three astronauts to practically zero. In the end, the astronauts would not even have had to be inside the Command Module when it was jettisoned, but could have boarded it only after the ditching.

What do you think, dear readers, which version should be in the history books? I have retold the story to you on the basis of the NASA documentation and given the sources for it. I have justified my judgments and interpretations, so that you can understand my trains of thought and also check, what I would like to encourage you to do. The times when you blindly believe everything in the media or believe that the majority is always right should be over. Maybe some sections were too technical for you, for example the back start from the Moon to the rendezvous. But in this case, you can also just compare what could be done in the 1960s and what can be done today, and how have the tools evolved. You can also check the credibility of space agencies by comparing: what have they promised and what have they carried out in the last 30 years. This is the basis on which many managers judge and decide. Do it the same way.

But the story is not finished yet. In the next chapter, let's look at how the astronauts fared after the Apollo 11 mission. From now on, I will increasingly rely on sources other than NASA.

5 Glorious Times

The astronauts spend the first 2 ½ days after landing in the quarantine container [93] and then in the Lunar Receiving Laboratory for a total of about 21 days in quarantine. [94][95] On August 10, 1969, the quarantine is lifted. [96]

On Tuesday, August 12, 1969, at 10 a.m., the three astronauts give their first press conference in Houston, the so-called post-flight press conference. [97] Fig. 79 shows a picture from the video of this press conference, where at the beginning just the three "heroes" are introduced - or don't they rather make the impression of three defendants who feel guilty? It looks like they have just been shown the consequences of what happened to them if they made a false statement.

The official celebration with parades through New York, Chicago and Los Angeles then takes place on Wednesday, August 13, 1969. On the same evening, the three astronauts are invited to the State Dinner in Los Angeles, where everyone who is anyone in the USA is present. This is followed by a 45-day tour of 25 countries.

Fig. 79 Aldrin, Armstrong and Collins at the Post-Flight Press Conference [97]

On Tuesday, September 16, 1969, exactly two months after the launch of Apollo 11, they speak before a joint session of Congress and present both chambers with a U.S. flag each that they have taken to the Moon. [96]

The astronauts have many joint appearances during this phase, but as time goes on, they go their own ways more and more often.

Michael Collins, the pilot of the Command Module, has the least publicity. He is not a Moon walker. But his history is all the more interesting: in 1968 he had disc problems in his cervical vertebrae, which affected his legs. He was operated and had to wear a neck brace for several months. After his recovery in late 1968, he was then nominated for Apollo 11. [98] This nomination fits the thesis that astronauts were grounded. To nominate someone for the most important of all missions, where accelerations up to 6.5 g are expected, who has just recovered from a back operation, does not seem real to me.

Edwin "Buzz" Aldrin is the most active when it comes to public relations. He has a doctorate in astronautics and is also an honorary doctor of science. Further he had already before his Moon flight the 32nd degree of the Freemasons [28] and in the same year he even received the 33rd degree. In 1971 he leaves NASA and soon falls into a depression, intensified by drug and alcohol abuse. In 1978 he overcomes his alcohol addiction and writes five books, appears in television series and films and gives lectures. He also appears on television as a space expert and advises companies on film productions. [99] In March 2016, he is coming to the Swiss mountains for an advertising campaign. Switzerland Tourism invited him and he accepted rather quickly. [100] In December 2016, at the age of 86, he makes a trip to the South Pole, which he is the oldest person to have reached. [99]

Reporter Bart Sibrel wants clear confirmation that Apollo 11 landed on the Moon, so he asks Buzz Aldrin to swear on the Bible that he walked on the Moon. He approaches Aldrin on the street, asks him this question and holds out a Bible to him. Aldrin evades, crosses to the other side of the street, but the reporter does not let up. When Aldrin continues to refuse to swear, the reporter calls him a coward and a liar, whereupon Aldrin punches him in the face. The scene is filmed and can be viewed on YouTube. Bart Sibrel published the video on February 18, 2016. [101][102]

Bart Sibrel also asked other "Moon walkers" to swear, but none took an oath that they had been to the Moon. [103]

Neil Armstrong shuns publicity and rarely gives interviews. Armstrong's career began as a Navy pilot, then he became a NASA test pilot and was selected as an astronaut in 1962 and assigned to the Apollo program also thanks to his courageous reaction during Gemini 8. After his Moon flight, he is promoted to deputy chief of NASA's Washington Aeronautics Office in 1970. In 1971, Armstrong leaves NASA, teaches as a professor of aerospace engineering until 1979, and then moves into business, holding board seats and management positions. From 1985 to 1986, Armstrong serves on the National Commission on Astronautics, and in 1986 is appointed second chair on the Commission to Investigate the Challenger Disaster. [104]

Fig. 80 Armstrong during his Speech on the Occasion of the 25th Anniversary [105]

Wednesday, July 20, 1994 marks the 25th anniversary of the Moon landing. The Apollo 11 crew is invited to the White House to see President Bill Clinton and Neil Armstrong gives a short speech (Fig. 80). In it he addresses a group of students who are also invited and tells them that only the beginning has been completed, much is left unfinished for them. Great ideas are still undiscovered, and breakthroughs could be made by those who remove one of the protective layers over the truth. To me, this sounds like an admission when he says here that the truth is hidden under protective layers. One can also interpret this statement as meaning that he wanted to give a sign to the world. I think he should not have gone further without this speech being censored. After all, he had to speak through the grapevine. By the way, he did not mention Apollo and his journey to the Moon with a word.[105]

On Monday July 20, 2009, the 40th anniversary of the Apollo 11 Moon landing, the three Apollo 11 astronauts are received by U.S. President Barack Obama. [219] At the press conference on the same day, only Buzz Aldrin of the three is invited; Neil Armstrong does not participate. [220]

On August 5, 2010, Neil Armstrong's 80th birthday, Austrian Servus TV shows a talk program with him that was recorded shortly before. The presenter introduces Armstrong and says that he has now done something that he had never done before in his life, as with the Moon landing, namely entering a television studio. Also invited is Alexei Leonov, the first person to step out of a spacecraft and Armstrong's friend. He gives Armstrong a lavish gift for his upcoming birthday. [106]

During the broadcast, the host then asks Neil Armstrong directly, "Mr. Armstrong, hand on heart, were you really on the Moon?" The astronaut looks surprised. The questions of the talk show had not been agreed upon beforehand. Armstrong remains silent. And remains silent. Then a broad grin stretches across his face and he says simply: "Definitely. Definitely."[107] According to another source, during the

Apollo 11 – The Real Story

silence Alexei Leonov comes to his aid and wakes Armstrong up with a laughing "Neil!", and a spectator shouts between them "Yes, it was him!". Only then Armstrong had answered with his double "Definitely". [108]

In a lecture hall, reporter Bart Sibrel asks Armstrong to swear that he was on the Moon, and offers him $5,000 in return - for himself or for a charitable organization. But at first Armstrong thinks the Bible is probably fake and at the end he says goodbye with the words that Sibrel does not deserve an answer. [103][109]

Neil Armstrong dies on August 25, 2012, as a result of bypass surgery at the age of 82. [104] If he had lived on until 2019, what do you think he would have said in his speech at the 50th anniversary?

However, the legend of Neil Armstrong lives on. Even after his death, he remains the most famous astronaut ever. Thus, at the beginning of November 2018, his two sons had 2000 mementos of their famous father auctioned off, generating proceeds of 7.45 million dollars. The most expensive piece, a plaque that Armstrong had with him in the Lunar Module, fetches almost half a million dollars. [110]

Apparently, his two sons wisely sold in time, before the publication of this book.

6 Manned Space Flight after Apollo 11

In the same year as the Apollo 11 Moon landing, a second Moon landing takes place with Apollo 12. The mission lasts from November 14 - 24, 1969 and follows the same scheme as Apollo 11. As with the Gemini programs, the U.S. demonstrates that it is capable of such missions. Apollo 12 lands within walking distance of Surveyor 3, which had landed there unmanned 2 ½ years earlier. (Fig. 5)

The Zürichsee-Zeitung headlines on Tuesday, November 25, 1969 on its front page "Apollo 12: New triumph, new records" with the subtitle "Lunar astronauts happily landed in the Pacific almost to the minute". The landing point on Earth had been only 5 km from the aircraft carrier; thus, after their precision landing on the Moon, the astronauts had also set a record for accuracy of navigation on their return to Earth. The Apollo 11 crew had landed 23 km from the carrier in July.

The entire undertaking was characterized by extreme precision. The scientists had achieved even more than they had had in mind. The astronauts had collected more stones and more carefully than anyone had ever hoped. Professor Conway Snyder, who worked on the scientific instruments, estimated that the Apollo 12 flight had revealed at least a thousand times, if not a million times, more information about the Earth's satellite than the July Moon landing. It was truly a giant leap, probably the biggest leap they would ever take toward understanding the Moon, he had said. - I comment on the mentioned benefit of the Moon landings in the next chapter.

Subsequently, the newspaper reports on some film and color television camera mishaps and that President Nixon has promoted the three naval officers who previously held the rank of frigate captains to sea captains.

Finally, the newspaper reports that shortly after the splashdown of Apollo 12, the Soviet news agency TASS had already reported on the successful completion of the American lunar landing enterprise.

This is the last article about the Apollo program that came to light while cleaning out my wife's parents' house. Apparently, the fascination with space travel passed its peak at the end of 1969. The news about Apollo 12 was placed first on the front page, but this time a scant half page had to suffice. In order to get more attention from the public and the media again, it would be helpful if the program were more varied. The lunar flight now following runs completely in this sense:

Apollo 11 – The Real Story

On April 11, 1970, at 13:13 Central Standard Time (Universal Time - 6 hours), Apollo 13 launches, like its two predecessors, from launch pad 39A at Cape Canaveral. The Command Module is named Odyssey, which also means errant flight. The landing of Apollo 13 is to take place in the Fra Mauro highlands, i.e. for the first time in a mountainous region. On the evening of April 13, when the crew is on its way to the moon, an oxygen tank in the Service Module explodes and also affects the other oxygen tank, causing the fuel cells and thus the power supply to slowly fail. The pilot of the Command Module John Swigert sees a warning light after the explosion and reports: „Houston, we've had a problem here".

Before the explosion, Apollo 13 has already made a course correction so that its course is aligned for a landing on the Moon, so they are no longer on a trajectory where they can return to Earth without propulsion. Without power, however, the Service Module's rocket motor cannot operate. Soon after, the bright idea comes: instead of the Service Module's engine, that of the Lunar Landing Module could be used to return to Earth; likewise, the Lunar Landing Module can provide power and oxygen. With the engine of the Lunar Landing Module, Apollo 13 thus brakes behind the Moon into a lunar orbit, makes one lunar orbit and then accelerates again with the same engine toward Earth. The Service Module is not detached until four hours before landing. Another three hours later, the crew climbs back from the Lunar Landing Module into the Command Module, detaches the Lunar Module and then lands in the Command Module in the Pacific Ocean on April 17, 1970. [111] Even after this ordeal and six days of weightlessness, the three astronauts easily manage to transfer from the wobbly Command Module to the inflatable boat and sit on the edge of the boat. [112] We will examine the astronauts' condition after landing on the Space Shuttle in more detail.

Between January 31, 1971 and December 19, 1972, four more lunar landing missions follow with Apollo 14 through 17, basically like Apollo 11 and 12. The outdoor missions become longer and the astronauts move farther and farther away from the Lunar Landing Module. At Apollo 14 they carry a two-wheeled handcart with them [113] and from Apollo 15 even a lunar car (LRV, Lunar Roving Vehicle). The Sun shines steeper and steeper, the steepest during the third outdoor mission of Apollo 16, between 45.8 and 48.7°. This mission lasts almost six hours. [67] The Sun's position is higher than at

Manned Space Flight after Apollo 11

noon in Zurich on March 21 at equinox. The lunar soil is already heated up to over 70°C at this time, almost 100°C hotter than during Apollo 11. [68] An air conditioner, which puts away 100°C temperature increase so loosely, is with this description probably also too much of the good.

As already mentioned in the introduction, I followed the television transmissions from Apollo 15 live; our family had acquired a color television set in the meantime. Above all because of the perfect picture quality I was never quite sure whether everything went with right things. I looked very closely, but could not find anything which spoke against a live transmission from the Moon. I also remember the hammer-feather experiment of Apollo 15, when commander David Scott dropped a hawk feather and a hammer side by side at the same time to demonstrate that in a vacuum both parts take the same time to reach the ground. [114] Today, this is often cited as proof that the astronauts were actually on the Moon, because on Earth the feather floated slowly to the ground due to its air resistance. At that time, I had not known how easy it would have been to recreate this scene on Earth: all you had to do was insert a metal pin into the quill and record the film 2½ times faster than you played it back later.

On April 19, 1971, with the Apollo Moon flights still underway, the Soviet Union launches the first Salyut 1 space station, followed by six more Salyut space stations until Salyut 7 burns up in the atmosphere on February 7, 1991. The Soviets thus gain their first experience in the long-term stay of cosmonauts in weightlessness. [115]

On April 22, 1971, the Soviet Soyuz 10 launches and completes the first docking with a space station. Soyuz spacecraft are permanently used until today first by the Soviet Union and then by Russia. The majority of the missions are shuttle flights to space stations or new spacecraft (e.g. Soyuz 12) are tested. The basic concept of the Soyuz spacecraft is always the same: in the back the instrument section with engine, fuel and oxygen tanks, and on which the solar panels are also mounted, in the middle the descent module with the cockpit, and in front the roundish orbit module or sojourn module with sleeping bags, toilet and storage space for material. Soyuz is adapted to its tasks and further developed. The TM series stands for flights to the Mir space station, TMA and MS for flights to the ISS. [116]

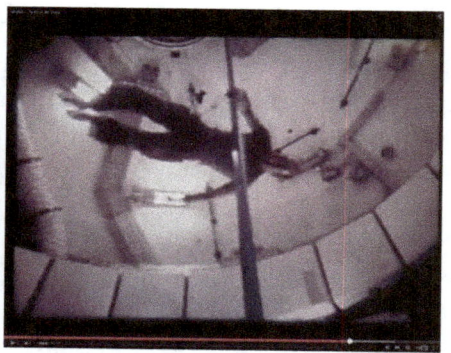

Fig. 81 Skylab [117]   *Fig. 82 Skylab 1, Exercise on High Bar [121] (≈ 18:00)*

The U.S. feels that it is in traction. They launch their first space station, Skylab, on May 14, 1973, using a Saturn V rocket. Since they wanted to recycle existing technology, they built Skylab from the third stage of the Saturn V rocket. During the launch of Skylab 1, a solar panel and the meteorite and thermal shields are torn off, causing the temperature in Skylab to rise as high as 52°C at times. To contain the excessive heating, a square solar shield is cut ad hoc. 11 days later (with a delay of 10 days) the crew starts with an Apollo Command and Service Module as Skylab 2 and reaches Skylab 1 in six hours. There the astronaut Paul Weitz tries for 40 minutes from the open exit hatch with a pair of cutting pliers mounted on a pole to loosen the second solar panel, which is jammed due to a foreign object, so that the panel could unfold. In the process, his colleague Joseph Kerwin holds him by the legs. But the operation is unsuccessful and Skylab 2 docks with Skylab 1, ending its first 22-hour day. The next morning, the astronauts go to Skylab 1, where it is 54 °C hot. They immediately install the square sunshade, which is visible on Fig. 81 as a golden sail. On it, the temperature drops to comfortable levels within the next three days. The crew can then perform the most amazing acrobatic tricks (Fig. 82 and Fig. 83) [117][118] [119][120]

Fig. 83 Skylab, Somersault [122] (≈ 3:20)

Manned Space Flight after Apollo 11

Above section about the Skylab could not be more implausible in my opinion. Apart from the fact that the time until the rendezvous and docking is much too short, as I have already explained in chapter 4.5, the whole story comes across as so cowboyish that it is unnecessary to go into detail about the individual sequences.

The whole thing culminates in a video where Skylab 1 is filmed from above with a fluttering solar sail. [123] Since there is nothing to flutter in vacuum, this video was recorded in air and not in space. The Skylab fits its name perfectly, as "sky" also means "air space".

NASA had called on students to make suggestions for space experiments. One of them was to check whether a spider could also build its web under weightlessness. This experiment was given to the second Skylab crew. After a period of habituation, the spider built a perfect web. [124]

Only now, as I write this respectively have just researched, I realize that the result from the spider web experiment, which I had previously classified as knowledge from space travel, is based on a fictional story and I must now admit to myself that I no longer know whether a spider could build its web under zero gravity and how long it would need to do so, since this experiment has never been done. This makes me almost a little sad, especially as the student who proposed this experiment was also exploited and lied to by NASA. Probably they chose her experiment because it can be easily replicated on Earth. After all, we can't tell from a photographed spider's web in which environment it was built.

On July 11, 1979, Skylab breaks up in the atmosphere and parts crash, having been visited by a total of three crews. [125] But as described above, I consider Skylab as pure staging.

On July 15, 1975, the Soviet Soyuz 19 launches from Baikonur and 7½ hours later the U.S. Apollo Command and Service Module CSM 111 launches from Florida to demonstrate a Soviet-U.S. rendezvous (Apollo-Soyuz Project). It takes them about two days to find each other and dock. [85] I have already described this maneuver in chapter 4.5 in the consideration of the rendezvous.

The two spacecraft separate again on July 19, with Soyuz 19 landing near Baikonur on July 21 and the Command Module (CM) landing in the Pacific Ocean on July 24. During landing, the CM goes into a spin and unburned gases enter the capsule. The splashdown is harsher than usual and the crew must remain in the hospital for two weeks for observation after recovery. This is the last flight of an Apollo spacecraft and also the Saturn IB. It is also the last manned flight in which the capsule lands in the water on parachutes. [126]

After many things were faked during the Apollo missions, I wonder if this was the first official flight of Apollo at all, where launch and landing took place manned. For this mission, the indication of time from launch to docking of about two days is in the range of what is needed even today. From this point of view, two days is a very good time. Since in today's comparison missions the target spacecraft always behaves passively, but in the Apollo-Soyuz project both spacecraft actively adjusted their position again and again, this mission could have taken place as described. This mission probably forms the transition from staged to real US missions. I think that the US missions described in the following have actually all flown manned or are still flying.

As a successor spacecraft to Apollo, the U.S. is building the Space Shuttle, a reusable spacecraft that launches like a rocket and lands on a runway like a glider (cf. Fig. 11 and Fig. 85). It can fly to low Earth orbits up to a maximum orbital altitude of 643 km. It has a large cargo bay, so it can also launch satellites in orbit. The first flight is on April 12, 1981 with the Shuttle Columbia and the last landing is on July 21, 2011 with the Shuttle Atlantis. [127] [128]

On June 29, 1995, the Space Shuttle Atlantis docks with the Russian space station Mir for the first time. [129] The last flight of a Space Shuttle to Mir takes place in June 1998. [130]

There are two accidents:
1) On January 28, 1986, during the launch of the Shuttle Challenger, a sealing ring of one of the solid rockets fails. A flame escaping from there destroys the hull of the propellant tank as well as the attachment of the solid rocket. The tank ruptures and the Shuttle is also destroyed as a result. All seven astronauts, two women and five men, are killed.[127] [131] Second chairman of the commission for the investigation of this catastrophe is - as I already wrote in the chapter 5 - Neil Armstrong.
This is the official story of Challenger. In 2015 a Dr. Eowyn publishes on a blog his investigations about this accident: six out of the seven crew members seem to be still alive. He found persons with an identical or a very similar name and approximately the same age. [222] What he publishes looks real to me. E.g. the pilot Michael J. Smith [223] seems to have a double with the same name. This Michael J. Smith is a Professor

Emeritus (retired) of Industrial and Systems Engineering at the University of Wisconsin-Madison. His picture is on the website of this university and shows the same striking features. [224]

2) During the launch of the Shuttle Columbia on January 16, 2003, some foam pieces break off

Fig. 84 Michael J. Smith: Astronaut 1986 and Professor Emeritus

from the external tank, possibly also pieces of ice. These punch a hole in the heat shield. Although the technicians in the control center notice the event, they are not aware of the damage caused. Then, during the return of the flight on February 1, 2003, hot plasma generated during re-entry enters the wing structure through the hole, the Space Shuttle breaks apart, and all seven astronauts, five men and two women die. [127] [132]

The Space Shuttle is very suitable to take a closer look again at the landing of a space flight and especially the exit of the astronauts. For this purpose, I found a YouTube video of the landing of the Shuttle Discovery on March 28, 2009. [133] Discovery launched on March 15, so the crew was in space under zero gravity for 13 days. All six male members are part of the crew and were in space for the 13 days described. Sandra Magnus, the only female, is returning from her four-month mission on the ISS. [134][135] Apollo 11 lasted 8 days and Apollo 13 lasted six days. So on the ISS, the stay was a little longer, with fitness equipment there to counteract the regression of muscles and bones. [136]

At the beginning of the video you see the descent and the spectacular touchdown on the runway. A mobile container then drives directly to the exit door and you don't see how the crew gets in there, i.e. whether the astronauts can walk themselves or whether they are carried. In the container they are first medically examined. Then, as is tradition, they will walk around the Shuttle. About an hour after landing, the six crew members descend the stairs of the container and walk with unsteady steps toward the NASA managers to receive congratulations from them.

Apollo 11 – The Real Story

Their gangly steps after 13 days in space and an hour of "resuscitation" seem authentic here, in stark contrast to the exits from the Apollo Command Module after the splashdowns, where the astronauts climbed out of the capsule on their own after a few minutes and, despite the swell, climbed safely and unassisted into the rubber dinghy and sometimes jumped. It is also fitting that Sandra Magnus is not yet able to walk around the Shuttle after her four-month mission on the ISS.

On November 28, 1983, Spacelab, a reusable space laboratory, launches on the Space Shuttle Columbia for a 10-day mission. It was developed and built by a consortium of European companies on behalf of the European Space Agency. It is designed to fit in the Space Shuttle's cargo bay, and it is used only in the Space Shuttle; thus, it never flies independently in space. The main component is a cylindrical pressure module 4 m in diameter and 7 m long. It has room for three science astronauts.

The last launch is on April 17, 1998, again with Columbia. This time the mission lasts 16 days.

In addition to the pressure module, the European Space Agency is also supplying three unmanned components:
1) U-shaped pallets that fit into the loading bay and on which instruments such as telescopes can be mounted
2) A tracking unit for telescopes as well as
3) An igloo; this is a pressure-regulated aluminum container in which the control system for the experiments of the pallets is packed. The igloo is used only during unmanned flights, that is, when the mission is carried out only with pallets.

The Space Shuttle's cargo bay is opened in space to operate the Spacelab and its pallets, so it flies with the roof open.

Unmanned missions using only pallets were conducted as early as November 1981, two years before the first Spacelab launch. The last unmanned mission was in May 2009, more than 10 years after the last Spacelab flight, and again about two years before the Space Shuttle was decommissioned. [137][138]

On February 19, 1986, the Soviet Union launches the base module of its new Mir space station in Baikonur. After its expansion, it is the largest artificial object in Earth orbit and is easily visible by eye at night. It remains in orbit until 2001, entering the atmosphere on March 23, 2001. Part of it burns up and the rest falls into the sea.

Manned Space Flight after Apollo 11

The first crew launches on March 13, 1986, and enters and commissions the space station two days later. As a special feature, a 50-day excursion is made to the Salyut 7 space station to maintain it and take over some of the equipment for Mir. This flight of a crew between two space stations is still unique today.

After the collapse of the Soviet Union, the Russian space agency Roskosmos has been operating the space station since 1992. As already described for the Space Shuttle, it flew to the Mir space station several times between 1995 and 1998.

Fig. 85 Space Shuttle Atlantis docked with Mir [141]

The orbital altitude of Mir is 390 km. Here the cosmic radiation is still moderate and also the braking effect by the upper atmosphere. [139][140]

On November 20, 1998, a Russian Proton rocket launches the first module "Zarya" of the International Space Station (ISS) into orbit from Baikonur. Since the U.S. had no experience in building propulsion modules for a space station, Zarya was designed and built on behalf of the U.S. in Russia. The cost of building Zarya was covered by the United States. About two weeks after its launch the STS-88 Space Shuttle mission transports the first U.S. module, the "Unity" connecting module for the ISS. Zarya and Unity are then coupled. By summer 2000, Zarya will provide complete power, attitude control and climate control for the space station. [142]

This formulation in Wikipedia fits perfectly with my assessment of Skylab: the U.S. could not, in fact, have gained any experience of operating a space station from the fabricated reports on Skylab.

Like Mir, the ISS has a modular design. Individual assemblies are brought into orbit by launch vehicles and Space Shuttles and assembled there. For example, in its tenth year of operation, the European research module Columbus will be installed on the ISS on February 11, 2008. It is visible on Fig. 86 on the front right.

Apollo 11 – The Real Story

Fig. 86 International Space Station ISS from above and behind [143]

When completed, the ISS will have a span and length of about 100 meters each and a depth of 30 meters. It is considered the most expensive object ever built by humans and the total cost is expected to be about 100 billion Euros. [144]

Like Mir, the ISS can be observed very well by eye in the night sky. It is usually brighter than the brightest star (Sirius), also brighter than Jupiter and almost as bright as Venus. A passage typically lasts five minutes. On the website of Heavens Above [145] you can see for any location when the ISS can be observed.

Since about 2010 one can observe on several so-called live conferences from the ISS that there are image errors. It is the kind of image error that is typically made by a computer which has to assemble two images: a foreground that is filmed in front of a green screen (in Fig. 87 with Samantha Cristoforetti) and the ISS-background. Other videos show gravity effects. This means that the named ISS astronauts remain on Earth and that the ISS is crewed by unknown astronauts and/or remote controlled.

In the movie "Gravity" (2013 by Alfonso Cuarón) Sandra Bullock moves through the Space Station as if it was real. Nobody assumed that it had been taken in space. Why should ISS live streams be recorded in space if the technology is available on the ground?

The altitude of the ISS rises from 350 km to 400 km or higher in 2012; with the higher altitude the NASA takes a higher radiation risk – possibly because it is now easy to exchange the astronauts without announcement. At 400 km the air resistance is smaller so that less fuel is consumed.

Another reason for not showing live pictures from the ISS might be the bad conditions inside the station. As in the MIR also in the ISS grow mold fungi and other microorganisms which look unappetizing and affect the health of the astronauts. NASA makes a better impression showing a well cleaned double of the ISS chambers in these videos.[226] [227]

Luckily the ISS will finally burn up in the Earth's atmosphere so that all traces will be erased.

Fig. 87 Samantha Cristoforetti in a so-called live video from the ISS [225]

On October 15, 2003, the People's Republic of China launches its first manned spacecraft, Shenzhou 5, on a 21-hour spaceflight using the Long March 2F launch vehicle. On board is the first taikonaut Yang Liwei. The capsule lands on a parachute in Inner Mongolia. The spacecraft's orbital module remains in orbit to conduct automated experiments. It burns up on May 30, 2004.[146]

On September 29, 2011, the first Chinese space station, Tiangong 1, launches aboard a Long March 2F launch vehicle. Tiangong 1 is 10 m long. It is first operated unmanned and the first to dock is the Shenzhou 8 unmanned transporter. On June 16, 2012, the manned Shenzhou 9 spacecraft launches, and its crew, two men and one woman, enter the laboratory two days later after successful docking. The crew will remain in the space station for 10 days before returning to Earth. Tiangong 1 will be manned again for 12 days in June 2013. After that, the space station will be used for Earth observation and research until March 2016. Then radio contact is lost, causing it to plunge uncontrollably into

Apollo 11 – The Real Story

Earth's atmosphere and break up on April 2, 2018. Parts that have not burned up fall into the Pacific Ocean.[147]

Tiangong 2, the second Chinese space station, has been in space since September 15, 2016. It has a length of 9 meters and a maximum diameter of 3.35 meters. A crew of three can stay on board for up to 20 days. On October 18, 2016, the Shenzhou 11 spacecraft docks. Two taikonauts form the first crew and stay on board until November 17, 2016. From April 22 to June 21, 2017, the unmanned space freighter Tianzhou 1 is docked. Among other things, it refills the space station's fuel tanks. As of the time I am writing this (2018), Tiangong 2 is unmanned.[148]

7 Benefits of the Apollo Moon Landings

7.1 General Benefit

At first I wondered whether a faked Apollo 11 mission would be of any use at all. But then I quickly concluded that all projects that are started in this world are useful to someone or at least someone believes that it is useful to him - otherwise one would leave the matter alone.

I take a differentiated approach when describing the benefits. On the one hand, the benefit depends strongly on the detailed knowledge one has of this mission. On the other hand, the benefit is different for different groups of the population. The masterminds and organizers of the Apollo 11 mission have a different benefit than the unsuspecting, gullible spectators think they have. And the part of the population, which recognized the deception and to which now also you should belong, judges its benefit again differently. Moreover, the benefit is subjective, which means that not all people with the same level of knowledge would describe their benefit in the same way.

I distinguish between the public or scientific benefit, which affects the entire world population, and the unofficial benefit or benefit of the organizers of this mission.

7.2 Public or Scientific Benefit

7.2.1 Overview of the Scientific Objectives

The science topics are described in the Apollo 11 Preliminary Science Report [149]. In the preface of 31 October 1969 it is stated that this is a preliminary report of the scientific observations, but that further results and detailed analyses are expected from the evaluation of the returned Moon rocks, the photos and the experiments.

In the table of contents, the photos come first. I have already mentioned in the chapter 4.4 *Moonwalks* that for the general public the photos are the most important and that NASA shows them until today again and again and symbolically. The photos in the first place in the scientific report confirm that NASA and I agree on this point. By the way, during an interview in 2018, NASA's Chief of Research Thomas Zurbuchen said that the goal of the Apollo missions had been to raise the U.S. flag on the Moon and leave footprints. And they had achieved that. [150] All the more reason to ask why it took six missions.

Apollo 11 – The Real Story

Shortly after the landing of Apollo 11, the two astronauts would have set up three scientific experiments: the seismometer (Passive Seismic Experiment Package PSEP), the laser reflectors (Laser Ranging Retroreflector LRRR) and the solar wind foil (Solar Wind Composition experiment SWC). These three experiments were part of the EASEP (Early Apollo Scientific Experiments Package). This package included further experiments which, in contrast to the three above, had to be actively performed by the astronauts.

I will now first give you an overview of the scientific observations and the experiments. I proceed according to the table of contents of the Preliminary Science Report. Subsequently, I will deal with two topics in more detail in a separate subchapter.

As described above, the photos come first. I have looked closely at some Apollo 11 photos in earlier chapters and debunked them as studio photos. Nevertheless, a separate subchapter follows below that goes into NASA's continuation of this story.

Chapter 2 is called "Crew Observations". This chapter was written by the three astronauts. In it they describe, for example, that they mostly used the recommended settings when taking pictures, but that they interpolated between different settings for some pictures on the lunar surface, and that they adjusted the focus in such a way that the object of interest was shown as well as possible. One recognizes this for example very well with the pictures of the footprints. Whether they would have seen stars, is not written in the report. To the visibility of stars I come back in the chapter 8 again. Further there is a section about the observation of the solar corona, but then this chapter ends and the following page is missing. Probably it is an empty page, because the article ends in the middle of the page. The next page, page 41, the first page of the next chapter, is also missing. There are pages missing elsewhere as well; this report seems careless as a result.

Chapter 3 describes the geological background of the Moon rock samples. It says where which rocks were found and individual rocks are presented and their nature is described. The photo with the footprint in the soft Moon soil may not be missing also. To this chapter the story comes to my mind that a curator of the Dutch National Museum noticed in August 2009 that an exhibited piece of Moon rock from the Apollo 11 mission was nothing else than a petrified piece of wood. US officials could not have explained this either. [151] Further it is to be mentioned to this topic that the University of Bern cooperated from the beginning in the Apollo program and represents this also today still as real, for

Benefits of the Apollo Moon Landings

which I am ashamed as Swiss and apologize in all form, as far as this is entitled to me. [152] [153]

Chapter 4 describes the soil mechanics. It is about whether the ground is stable enough for a safe landing of the Lunar Module and whether it is possible to walk on the ground without sinking. This chapter also deals with the question to what extent fine-grained and statically charged sand adheres to the astronauts' suits. As in the last chapter, the footprints in the lunar soil are again emphasized here.

Chapter 5 contains the preliminary results of the investigations of the returned Moon rocks. Here the composition of the lunar rocks and the age determination are described. The oldest returned lunar rocks could be older than the oldest lunar meteorites found on Earth so far. If these did not show yet the maximum age, then the hope is expressed that one will find Moon rocks in other regions, whose age reaches back to the formation of the Moon. So it is subtly suggested that for a better scientific evaluation probably further Moon missions are necessary.

Chapter 6 describes the seismometer respectively the possibility to register now Moonquakes and waves in the Moon surface. Measurements are already presented, which were caused by the astronauts or the back start of the ascent stage.

Chapter 7 is called LLRR, i.e. Lunar Laser Ranging Retroreflector. In this chapter already successful laser distance measurements to the Moon respectively to the retroreflectors installed there are described, of which the first ones would have taken place in the night of August 1, 1969. The results of these laser distance measurements from the Earth to the Moon are often considered as proof for the authenticity of this Moon mission. It is worth to have a closer look here. I dedicate a separate subchapter to this topic.

Chapter 8 describes the experiment with the solar sail that captures the solar wind. The solar sail is shown in Fig. 39 and also in Fig. 52. It is an aluminum foil, 30 cm wide and 1.4 m high, which is hanging on a pole and aligned to the Sun. I have already described the solar wind in chapter 4.1 in connection with the Van Allen radiation belt and cosmic rays. The protons, electrons and other charged particles emitted by the Sun remain hanging on the foil. The chapter goes on to say that the sail would be repacked before the return launch and that the composition of the solar wind could then be determined in a laboratory on Earth. The sail was sent to Bern, where it arrived on August 12, 1969 and was subsequently evaluated. Also for this inglorious cooperation of the University of Bern with NASA I would like to apologize as a Swiss, whereby I would like to add to the exoneration of the university that this experiment could have been faked very well, for example if NASA had unfolded this sail in space outside of a satellite and thus had captured the real solar wind or if it had been irradiated appropriately well in a laboratory.

Chapter 9 summarizes the results of the stereoscopic photographs. The stereo camera can be seen on Fig. 56. With this camera, the lunar soil could be photographed stereoscopically on an area of 7 cm x 8 cm.

Chapter 10 describes the dust detector. This detector is intended to measure the dust that is stirred up by the back start of the ascent stage or that accumulates there over a long period of time. The dust settles on solar cells and thus reduces their efficiency. At the same time, this experiment could be used to measure the reduction in solar cell efficiency during a solar storm.

The following two subchapters now deal with two so-called scientific findings in a little more detail:

7.2.2 Moon Images

The lunar images have been revealed in earlier chapters to be reenactments on Earth. However, the story of the landing site images did not end with the Apollo missions. NASA continued to spin the thread by releasing more images of the landing sites in 2009. These images were taken by the Lunar Reconnaissance Orbiter LRO, a NASA satellite that launched in June 2009, orbited the Moon for an extended period of time, photographed and mapped [154]. Thus, NASA was able to publish images of the landing site on the 40th anniversary of the Apollo 11 Moon landing and thus substantiate the credibility of this mission.

Fig. 88 shows the landing site of Apollo 11, taken by LRO between July 11 and 15, 2009. The image width is 282 m. You can see the lower stage of the Lunar Landing Module and its shadow.

Fig. 88 Landing Site of Apollo 11 ([155], Arrow from Author)

Benefits of the Apollo Moon Landings

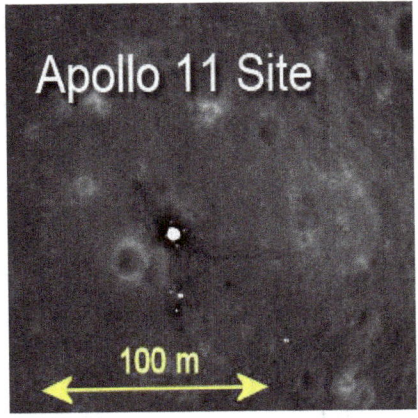

Fig. 89 Landing Site of Apollo 11 ([156], Detail, Length given by the Author)

Fig. 89 is another image of the Apollo 11 landing site. This picture was taken on November 9, 2009, around noon. Therefore no shadows are visible. The upper part of the lower stage glows in blinding white. I would have expected a dark tone, since it should be scorched by the return beam of the launching upper stage.

In general, I judge the quality of these pictures as very poor. Besides, with today's image processing software almost everything can be done, so that pictures alone are hardly conclusive today. With the knowledge that the Apollo 11 Moon pictures are studio pictures, NASA disqualifies itself again with these publications.

7.2.3 Laser Distance Measurements to the Moon

Fig. 90 Laser Retroreflector from Apollo 11 (AS11-40-5952.jpg)

The first laser distance measurement was made in 1962, seven years before Apollo 11. In such a measurement from the Earth to the Moon, the time of flight of a laser pulse is measured. The laser pulse leaves a telescope pointed at the Moon. It then flies to the Moon as a cloud of light or photons, where it is scattered by the lunar surface in all directions. Some of the backscattered light is then received back by the telescope where it is directed to a detector. This detector is able to detect even single photons. So one measures the time from the start to the reception of the laser pulse respectively from what still comes back. Since the laser light propagates at the speed of light, i.e. at 300,000 km/s, the distance can

Apollo 11 – The Real Story

be calculated directly from the time of flight. The scattering of the laser pulse on the lunar surface works in the same way as when a laser pointer is pointed at a wall. Also from there the light is scattered in all directions: all observers can see the point well, independent of their location.

The Moon is 380,000 km away from the Earth. So the light or the pulse needs 2.5 s for the way from the Earth to the Moon and back.

During the Apollo 11 Moon landing, the astronauts are said to have left behind a laser retroreflector which, like a cat's eye, throws the light back in the same direction from which it came. Fig. 90 shows this reflector. It is supposed to be aligned with the Earth. Although the light is always reflected back in the direction of reception, the return reflection is greatest when the beam is perpendicular to the retroreflector.

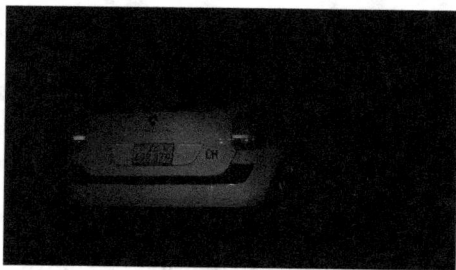

Fig. 91 My Car diagonally from behind

To better illustrate the principle of the cat's eye, I took two pictures of my car. I illuminated my car diagonally from behind with the headlights of a second car. Then I first sat down in front of the rear car between the two headlights and I made the picture as shown in Fig. 91. The viewing direction is the same as the lighting direction. The cat's eyes shine brightly and are visible as bright stripes because they reflect the incident light back in the direction of illumination. Then, with the same arrangement, I photographed my car straight on from behind (Fig. 92). The cat's eyes or retroreflectors are not shining now, because they reflect the light only in illumination direction, in Fig. 92 they reflect diagonally to the right rear.

Fig. 92 My Car straight from behind

The laser retroreflector works in the same way as the cat's eye of a car, it reflects the incident light back in the direction of illumination.

Benefits of the Apollo Moon Landings

This creates a bright light reflection. The laser retroreflector on the Moon is a cat's eye on the dark lunar background; the Moon reflects back only one tenth of the incident light. The lunar surface is about as bright as dark gray asphalt. With such a dark background in Fig. 91 you would see essentially only the back reflection and in Fig. 92 everything would be dark gray.

In the laser distance measurement to the Moon, the transmitting beam is very strongly bundled, so that it illuminates only an area of about one kilometer in diameter on the Moon. The location of the retroreflector is therefore not found at first go, but one has to make a search movement and then detects a 1000-fold increased return signal when one hits the retroreflector. Then one stops the search movement and makes a series of distance measurements.

Thanks to retroreflectors, it is guaranteed for later measurements that one measures the same point of the Moon in order to be able to estimate whether the distance changes with time.

Allegedly the measurements on retroreflectors have been carried out since August 1969. Apollo 14 has placed a retroreflector of the same type as Apollo 11 and Apollo 15 has placed a retroreflector three times as large, which should therefore also produce three times more reflected light. Measurements to these reflectors were made again and again until today and they should work perfectly since the beginning. [157]

If one has noticed now that Apollo 11 is based on fantasy, then the question arises what was measured since when. It could be that retroreflectors were placed on the Moon in another way and that these could then be used. But placing a laser retroreflector with an unmanned mission is not easy. Just landing with a rocket engine ahead on a large celestial body is very challenging; on Earth, this has only worked since 2015, as I described in chapter 4.3.

I searched for reports of different laser distance measurements and found some by which I could judge the quality of the measurement. The result is sobering. None of the three observatories examined measured a retroreflector-enhanced return. All three received only as much light from the Moon as one receives when illuminating the bare lunar surface. In other words, it was not possible to take any series of measurements at all over a long period of time, because without a reference point you never know if you will measure the same place again. [158]

The estimation of how much light comes back from the Moon was not very easy and is difficult to understand for a layman. By the way, this is also impossible for most astrophysicists without extensive

Apollo 11 – The Real Story

research. I learned this when I was looking for someone to give me the OK for my publication. Some astrophysicists within this field waved it off, saying they had no expertise in Lunar Laser Ranging. But if you want to check this against the reference given, you can still do so using the method of comparison. I calculated how much light should come back for both cases, scattering from the lunar surface and reflection from the retro-reflector. In the case of the retro-reflector, my estimate agrees with that from an observatory, and in the case of the scattering, the agreement with the 1962 measurement is good. So these two comparisons confirm my calculations. And all the measurements made by the observatories perfectly match laser pulses scattered from the lunar surface. By the way, only one of the three observatories considered was so open to state how much light should have been received from a retro-reflector on the Moon.

From these measurements one can conclude that there are no retroreflectors on the Moon and that the result reported by the scientists that the Moon moves away from the Earth by 3.8 cm per year is fictitious. (See also §10.8 of the Appendix) There is no reference point on the Moon due to non-existing reflectors, to which one can refer over a long time. However, the result sounds good. Sometimes a tolerance of $\pm\,0.07$ cm/year is given in addition, so that the layman and also the scientists are still more astonished. The distance rate is so small that this cannot be checked in other way in the next decades. Besides, a slow increase of the distance Earth-Moon is reassuring, one does not need to worry that the Moon will fall soon on the Earth.

The statements that the Moon moves away from the Earth with a speed of 3.82 ± 0.07 cm/year and that the retroreflectors functioned perfectly since the beginning together with a completely wrong estimation about the expected reflected light from the Moon I found by the way in an "Invited Review Article submitted to *Science*", i.e. in a publication in one of the most renowned scientific journals, and even in an article to which the author was invited by the journal. [157] If one knows that it is the highest of feelings for every physicist to be allowed to publish in the journal *Science*, let alone to be invited to do so, then I wonder who is pulling the strings behind this journal and controls what is published and what is not.

Obviously, NASA seems to have a direct line not only to the mass media, but also to the scientific publication organs. Over this wire it can publish apparently what it wants, and it exercises additionally also directly or indirectly control over what can be published.

7.2.4 Summary of the Public Benefit

What do we know more or better today thanks to the Apollo missions than we did before 1969? Did these missions bring any benefit?

The invented stories about the Apollo Moon landings did not bring any new information about the Moon at all. They could not do this. The scientific benefit is therefore exactly zero for those who recognized the falsification. Earlier I knew that I did not know how the Moon surface looks like, and I knew that the Moon always draws approximately the same circles respectively ellipses around the Earth, and I was also aware that the Moon orbit like all orbits of celestial bodies can change in the course of time, but that I did not know to what extent this happens.

For the gullible, the scientific benefit is negative. Today, many believe they know exactly what the lunar surface looks like, and they also think they know the behavior of the lunar orbit to within a few centimeters. So they have replaced their ignorance with false knowledge. The scientific benefit for these people is therefore less than zero or overall negative.

Since science considers the Moon landings to be real, the Apollo missions brought a scientific step backward regarding the Moon and also regarding expectations for manned spaceflight.

If NASA has been spreading false information about the Moon for the majority of the time since 1969, I wonder how they handle it with other celestial bodies, for example, whether the pictures of Mars really come from Mars. In any case, NASA could save a lot of money with fakes. If studio images are accepted for the Moon, why should NASA bother to produce real images for Mars, which would be much more expensive, and which it might not even want to show us?

In July 2018, NASA announced its Mars rover *Opportunity* had now been operating for 15 years. The plan was for 90 days. [159] Looking at the pictures, the tracks in the sand seem to be the most important thing here, too. To this I would just like to add that much of spaceflight is not scientific, as it is neither traceable nor independently verifiable.

Personally, the fake Apollo Moon landings opened my eyes to disinformation and global manipulation. The documentation is publicly available, the archives do not need to be opened further. A public confession of NASA or whoever is not needed in my opinion. Just the pictures recreated in the studio speak volumes. The more people realize this, the more aware the population will become and the more difficult it will be to manipulatively steer them in the desired direction. So the public benefit in terms of knowing about global disinformation and manipulation is positive. The more people realize this, the greater this benefit becomes. I think we can learn a lot from this for other areas of life in this day and age.

7.3 Unofficial and Secret Use

7.3.1 NASA as Top Dog

The majority of the scientific data, Moon maps and Moon pictures come from NASA or NASA-related US organizations. This comes once from the fact that the USA together with the Soviet Union were the two first space nations and that the USA remained also with a large budget at the topic. When at the end of the 20th century and at the beginning of the 21st century also Europe, Japan and China became fledged in space technology, the Moon remained however to a large extent in the hands of the USA.

In August 2013, NASA published a list of lunar missions based on its LADEE lunar mission under the title *Historical Exploration of the Moon* [160]. The first mission mentioned is Luna 2, which was the first to reach the moon in 1959 and crashed there. Then comes Ranger 7, which was the first to provide close-range images of the lunar surface. As described in chapter 2, Ranger 7 had taken these images just before it hit the Moon. Then comes the first soft landing of Luna 9 on the Moon in 1966, followed by five U.S. missions including Apollo between 1967 and 1972. In 1976, Luna 24 follows as the last mission to bring lunar rocks to Earth. This was followed by three NASA missions in 1990, 1994 and 1998, launches of Japanese, Chinese and Indian satellites to the Moon in 2007 and 2008, and finally two NASA missions, the Lunar Reconnaissance Orbiter LRO and another satellite in 2009, and GRAIL, a satellite for measuring the gravitational field of the Moon, in 2011. The satellite-based scientific measurement data on the Moon comes from NASA. Even if one searches further at Wikipedia [161] this impression is strengthened. The European space agency ESA has carried out a single Moon mission with SMART-1 in 2003. Countries

Benefits of the Apollo Moon Landings

like Japan, China and India sent individual probes to the Moon, but hardly any new data or knowledge was generated.

According to international law, all nations have free access to outer space including the Moon. More than 100 countries of the world, including Germany, Austria and Switzerland, have signed the Outer Space Treaty. [162] Article I states that there is unrestricted access to all areas of the Moon. Nevertheless, de facto NASA seems to have the Moon for itself. No scientist dares to propose a lunar mission these days. In scientific circles it is said that the Moon is no longer interesting for further investigations, because everything has already been investigated. If a scientist wants to receive money for a mission today, he must propose something new, something that no one has done before. According to the scientific consensus, this is hardly possible with the Moon, although, as we have seen, the lunar surface alone is anything but explored from close up.

Even if a scientist had recognized that the results published by NASA are from the fantasy, he would have nevertheless no chance to receive money for an investigation for example of the Moon surface, since he would have to convince various committees of his project. Soon he would be considered a conspiracy theorist and thus endanger his own reputation and career. So space scientists are looking elsewhere; the European Space Agency (ESA) has conducted a mission to Saturn and its Moon Titan (Cassini-Huygens, 1997-2017) and has a probe at Mars (Trace Gas Orbiter; launched in 2016) to study trace gases. In October 2018, BepiColombo was launched with Mercury as its destination, where it is expected to arrive in 2025. Further, ESA has a mission to Jupiter (JUICE, starting in 2022) in preparation. There are only vague plans for the Moon [163]; a landing in a crater near the South Pole planned for 2018 was stopped in 2012. [164] (As of 2018)

A small example of how NASA handles possible lunar material within the U.S. was recently provided by a woman who had received a vial of Moon dust from Neil Armstrong in her teenage years. This vial had recently resurfaced, she said, and now she was prophylactically suing NASA to make sure it did not take the Moon dust away from her, as NASA had done in previous similar cases. By the way, an expert had tested the Moon dust for authenticity. The result was double-edged. According to a first test, the composition corresponded to that of the known Moon dust, but according to a second test, the contents were typical Earth dust. [165]

In chapter 4.4.8 I mentioned the reaction of the Zürichsee-Zeitung to the Apollo 11 Moon landing. The editor-in-chief said there that the whole world had a share in this flight and also in the Moon and expressed his relief that this step had not succeeded to the Soviet Union, with which the fear existed to seize the Moon for their own purposes.

Isn't that exactly what NASA has done? It looks like NASA has done just that. It has staked out its territory from the beginning with symbols such as the footprint or the astronauts saluting the flag and has subsequently maintained this with many missions and corresponding publications. According to the Outer Space Treaty, all nations have access to the Moon, but the situation among scientists is such that hardly anyone dares to propose a Moon mission.

7.3.2 Technology Development

Before and during the Apollo program, NASA was able to develop technology for space travel. Infrastructure was built on the ground with control centers and powerful antenna systems, rocket engines, navigation systems, heat shields and much more was developed that was used after Apollo. NASA as a large organization was able to build and fly the Space Shuttle after Apollo.

Of course, there were also some curiosities that NASA developed or rather prototyped only to support the Apollo Moon landings, such as the Lunar Module and later the Moon rover. Such components served as a backdrop, as in a theater performance, but only made up a small proportion of the total budget.

Under the guise of the Apollo missions, NASA was able to become one of the leading space organizations and thus realize more and more real space projects. The budget could not have been larger. In addition, NASA may have developed technologies for other ventures that we do not know about. I will discuss this in the next chapter.

7.3.3 Investment in Secret Projects

NASA's budget for the Apollo missions was so huge that the question arises whether NASA did not carry out other projects in secret in addition to the visible technology developments. The facts are naturally thin on such topics, but I would still like to address this issue here. I refer to two book sources, for the time being to the book *Human Race Get Off Your Knees: The Lion Sleeps No More* by David Icke [166]. I have read a number of books by David Icke. He has done a lot of good

Benefits of the Apollo Moon Landings

research. For example, in his book *Alice in Wonderland and the World Trade Center Disaster: Why the Official Story of 9/11 is a Monumental Lie,* as early as 2002, a year after the attacks in New York, he revealed some things that could not be true about the official version, pointed out connections and asked good questions. In the book *...The Lion Sleeps No More* there is a section *Fake Moon Landing*, where he writes on the one hand that Kubrick produced the Moon landing in 1969 in a studio, on the other hand he reports that Kennedy shortly after his election got to see the technology on which the flying saucers were based. Shortly thereafter, he says, he gave his famous speech calling on NASA to send a man to the Moon before the decade was out. This could have been a dodge to get NASA to publicize its saucer technology. In other words, this means that NASA knew or was developing technologies such as antigravity "UFOs" at that time and Kennedy wanted to make this public. Obviously Kennedy did not succeed in this. He was assassinated before and NASA faked the Moon landing with conventional rocket technology. If antigravity drives really exist, they could explain various phenomena. For example, UFO observations are described as the UFOs making very fast, almost jerky motions [167] which fit perfectly to drives which can switch off the gravity. Antigravity technology also fits the structures built of large, heavy stones. The Egyptian pyramids of Giza are the most famous example, but in Peru, for example, there are also many huge stones. In the wall of Sacsayhuamán near Cusco, the heaviest stone weighs 200 tons. [168] In Baalbek in Lebanon there are worked monoliths from 1000 to 1650 tons. [169] A technology for the abolition or for the reduction of the gravity or levitation would have simplified much with these buildings or perhaps even made possible first.

David Icke mentions in connection with the faked Moon landing MK ULTRA, a mind control program of the US government. This program, like the antigravity technology, was also initiated by so-called paperclip scientists, i.e. scientists who were brought from Germany to the USA after the end of the Second World War by the operation "Paperclip" and continued their work there. This leads me to the second book source (*The TranceFormation of America,* [170]), which describes an apparently true-life story of a CIA slave under MK ULTRA, more precisely under Project Monarch. Under Monarch, previously abused children were recruited for trauma-based mind control operations. They were raped and tortured so that memories could be split off - memories can also be split off in normal life, for example by an accident, after which victims also cannot remember how it happened. Thus, these youngsters could be used to deliver secret messages. By erasing the memories, all traces could be removed. Hillary Clinton also plays a role in this book, a small one; nevertheless, I was relieved when she lost the presidential election in the fall of 2016. One chapter in this book is *Mind Control Training in the U.S. Military and NASA*. There was apparently a mind control lab in the basements of NASA Goddard Space Flight Center. The proximity to NASA also allowed access to other new technological developments, for example, sensory deception devices. It is interesting in this context that the first film Stanley Kubrick released after 1969 is about Mind Control. The film is called *A Clockwork Orange* and dates from 1971.

These two topics show what else NASA could have done with the money it received. For me this sounds credible, but as already mentioned at the beginning of this chapter the evidence is difficult. To UFOs I come in the chapters 8.3 and 8.4 to speak again. Mind control seems to me a topic on which the powerful of this Earth would have to rush whenever possible, in order to come still to more power. Mind control is probably the most powerful control of all.

The money for space travel flowed so lavishly in the USA in the 1960s also because it could be said that one was in competition with the Soviet Union, and that the latter was additionally a military threat. The same was true in reverse for the Soviet Union. Thus, space and other industries on both sides could benefit from this competitive situation. The budget of NASA can be seen in Fig. 93; the most impressive curve is the top one, where the budget is shown relative to the value of the dollar in 2014. In 1965-1967, annual budgets were twice what they are today. The middle curve shows the NASA budget in absolute U.S. dollars. As a result of the devaluation of money, it is not possible to make comparisons with this curve; that is what the upper curve is made for. These two curves necessarily intersect in 2014, since the dollar has the same value in absolute terms and in relation to 2014. The lower curve again represents the NASA budget, this time as a percentage of the US budget. In 1965 and 1966 this exceeded the 4% mark. [171] For comparison, I have also plotted the 2019 space budget of the US Air Force. This is similar in size to NASA's, with an estimated 20 billion US dollars. This budget is not directly accessible. It is made up of a non-classified budget of about $8.5 billion and a classified or secret portion. This secret portion is said to make up the largest piece of the secret Pentagon budget of $48.7 billion. [172] If I assume $11.5 billion for the

Air Force's secret space budget, together with the unclassified $8.5 billion, I arrive at the $20 billion given above. So the largest space budget in the U.S. is not necessarily NASA's, but more likely the one of the U.S. military, whose budget is 25% larger than the one of the Air Force, which receives 80% of the space military budget. [173]

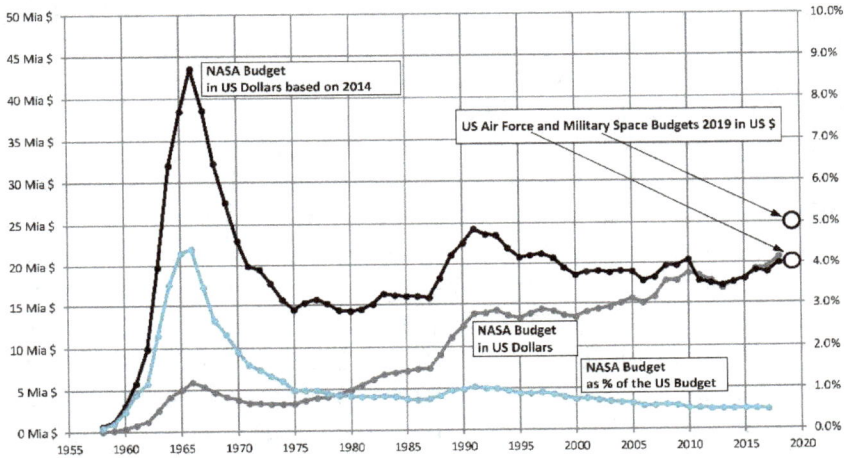

Fig. 93 NASA Budget [171] and estimated 2019 U.S. Military Space Budget

This chapter is intended as food for thought for all those who are hearing about the topics addressed for the first time. Stay critical, get more information and read more books about it.

8 Moon Landing, Conspiracy Theories and Wikipedia

8.1 General Conspiracies

Now follows a rather philosophical chapter, where I go into the question whether improper or manipulative information dissemination also occurs with other topics than the Moon landing. Furthermore, I show that there is a strict control at Wikipedia, so that examples like I list here cannot be published there.

On the subject of the Moon landing, as with other sensitive topics, the word "conspiracy theory" often comes into play. This term is used to nip counter-arguments in the bud. Only the ending "-theory" suggests that it is only a pure possibility, that the reality is however another. Thereby conspiracies occur frequently in the daily life. In a conspiracy, at least two people forge a secret plan and then carry it out undetected. The public does not find out who is behind this action; those involved maintain silence. A famous conspiracy is, for example, the September 11, 2001 attacks in New York. Since no one has claimed responsibility for these attacks, there is a conspiracy. It is unclear whether Bin Laden conspired with his henchmen or whether some other organization conspired. But a conspiracy it is in any case. In this example, however, the word "conspiracy theorist" is applied only to those who question the official version. The word "conspiracy theory" has extremely negative connotations. It is used as a power phrase with the meaning "stop thinking for yourself, but obligingly believe the official version". There is also an effort to invent the most simple-minded arguments possible that the alleged "conspiracy theorists" would put forward in order to challenge the official version. More of this in the next chapter.

The Moon landing is often taken as a prime example of an event that should not be doubted. For example, when I bring up topics associated with the term "conspiracy theory," I often get the retort, "I guess you don't believe in the Moon landing either." This points out how simple-minded the argument often is. But the inventors of the power phrase "conspiracy theory" are pleased. They have achieved what they aimed for, namely that all topics with the label "conspiracy theory" are lumped together and that the official versions are believed.

One may conclude in no case on the inversion, that is, if now with the Moon landing was lied, that also with all other "conspiracy"-topics only lies are told. But the realization that some manned space programs including Apollo are faked encouraged me to take a closer look

Moon Landing, Conspiracy Theories and Wikipedia

elsewhere as well. And I can assure you, Apollo is not an isolated case. If I had concluded that while NASA faked the Moon landings, we were otherwise geopolitically correct in our information, I would not have written this book. After all, the Moon landing could be classified as a boy's prank. It made space travel known and popular, many were fascinated, were proud and rejoiced. I, too, was fascinated by all the technical possibilities and thus ended up working for a space company. I have many colleagues who would be very disappointed if they realized that all the books and brochures they received about the Moon landing were only based on fantasy. But because I have always had an urge to find the truth, I pursued what little circumstantial evidence I had at first. However, I made the decision to write a book only after I realized that the Moon landing is absolutely not an isolated case, but rather the rule of geopolitical information. The falsification is absolutely obvious from the technical point of view, almost as obvious as the kind of the collapse of the WTC 7 (I still come to it), and therefore especially for technically interested people best suitable to notice at all once in which mass is lied on this Earth.

Denouncing the USA or NASA is only a side effect of this book. My main goal is to sensitize my readers. The fake Moon landing should also be a door opener for you to other delicate topics and it should help you that you are not (anymore) so easily manipulated. I will still address the most important topics known to me. But I am aware that I may not yet have recognized other and more important topics myself. Therefore, I am curious whether you, dear reader, in particular, will be encouraged by this book to provide enlightenment in another area.

Conspiracies are nothing new. On April 27, 1961, President John F. Kennedy gave a speech to the American Newspaper Publishers Association at the Waldorf-Astoria Hotel in New York. I quote from his speech [174][175]:

„It requires a change in outlook, a change in tactics, a change in missions -- by the government, by the people, by every businessman or labor leader, and by every newspaper. For we are opposed around the world by a monolithic and ruthless conspiracy that relies primarily on covert means for expanding its sphere of influence -- on infiltration instead of invasion, on subversion instead of elections, on intimidation instead of free choice, on guerrillas by night instead of armies by day. It is a system which has conscripted vast human and material resources into the building of a tightly knit, highly efficient machine that combines

Apollo 11 – The Real Story

military, diplomatic, intelligence, economic, scientific and political operations.

Its preparations are concealed, not published. Its mistakes are buried, not headlined. Its dissenters are silenced, not praised. No expenditure is questioned, no rumor is printed, no secret is revealed. It conducts the Cold War, in short, with a war-time discipline no democracy would ever hope or wish to match.

... I am not asking your newspapers to support the Administration, but I am asking your help in the tremendous task of informing and alerting the American people. For I have complete confidence in the response and dedication of our citizens whenever they are fully informed.

... confident that with your help man will be what he was born to be: free and independent."

This speech seems to me more relevant than ever. At first glance it is not tailored to the Moon landing, perhaps more to the attacks of September 11, 2001 in New York. But at second glance, the Moon landing also fits perfectly into this scheme. Just think of the unofficial and secret use described in the chapter 7.3. Moreover, there are certainly many human fates involved in these space programs. In all probability, people who did not cooperate had to give up their lives. Stanley Kubrick has hidden his work for Apollo 11 in the movie "The Shining". The way he depicts the two twins with huge streams of blood, this seems to be an allusion that a lot of blood also flowed during the Gemini program. Other allusions in this film include a forbidden hotel room that is number 231 in the book. Kubrick changed the number to 237. 237 matches the Earth-Moon distance, which is 237,000 miles. In a key scene, Danny, the son of lead actor Jack Torrance, is playing on a hexagonal patterned carpet reminiscent of the launch pads at Cape Canaveral. When Danny stands up, you can see a rocket labeled "Apollo 11 USA" on his sweater. [176] Last but not least, the main character Jack keeps sitting at his typewriter to write a play. When his wife Wendy once manages to take a look at one of the many sheets of paper, she sees that her husband has filled all the sheets with the same sentence "All work but not play makes Jack a dull boy".[177] What is even more piquant about this sentence is that, as was customary in the past, the small L and the number one were written the same way on the typewriter used. So instead of "All work ..." you can also read "A11 work ...", and A11 is the abbreviation of Apollo 11.

Moon Landing, Conspiracy Theories and Wikipedia

The last film by Stanley Kubrick (7/26/1928-3/7/1999), Eyes Wide Shut, is about a Satanic mind control network of elite personalities and families. The producer apparently wanted Kubrick to recut the film, but he refused. Kubrick is said to have negotiated lifelong freedom from censorship for his involvement in Apollo. But as if on cue, he then died of a "heart attack", and about 15 minutes were cut from "Eyes Wide Shut" (a trigger phrase from mind control) before the film's release. The film was released, as Kubrick had contracted, on July 16, 1999 - exactly 30 years to the day after the launch of Apollo 11. [166]

The mass media, i.e. most of the big newspapers, most of the big TV stations and many information sites on the Internet are subject to the main stream, i.e. they take over unseen the reports of the big news agencies, to which also NASA has a direct access. Otherwise, the faked Moon landing, the staging of September 11, 2001, and much more would have already become part of general knowledge. (Still) it is possible to publish such information in book form. Help that this remains so or even becomes better again. Orson Welles is said to have caused a mass panic on 10/30/1938 with the radio play "War of the Worlds", [178] because the radio listeners believed in a genuine report about an attack of extraterrestrials. After all, at that time, people were able to correctly assess the situation within a day. In the case of the Apollo 11 Moon landing, the official misjudgment in the Western world has lasted for over 50 years.

8.2 How does the Manipulation

By manipulation I understand in this book an influencing or purposeful steering. This influence is usually done in such a way that the manipulated persons do not even notice that they are being manipulated. People believe they are deciding based on objective information. As we have seen from the example of the Apollo 11 Moon landings, this information is anything but objective. For example, in Chapter 5 of the Apollo 11 Preliminary Science Report, it is suggested that subsequent lunar missions might find older lunar rocks, with ages going back to the formation of the Moon. Based on such information, politicians then approved money for further lunar missions with a clear conscience and the population went along.

The Apollo Moon landings have been a recurring presence in the media. The media effectiveness is confirmed, for example, by the Switzerland Tourism's advertising campaign with Buzz Aldrin addressed in chapter 5. NASA also repeatedly ensures that the Moon landings remain in our consciousness. For example, the original photographs are suddenly untraceable and then they reappear. [179][180] Also the LRO pictures for the 40th anniversary showed who is master on the Moon. Our media print all messages of NASA dutifully. Universities and scientists also like to join in or work hand in hand with the media. It is generally the case that the media like to obtain expert opinions on certain topics. These experts are very often university professors who, if they want to keep their jobs, have to go with the flow. A counter-example is the Swiss historian Daniele Ganser, who dared to ask whether in the attacks of September 11, 2001 in New York, the third building in Manhattan, the WTC 7, whether this building had not collapsed as a result of an explosion when it collapsed at the speed of free fall. He still asks this question today, but had to give up becoming a university professor.

The permanent repetition of a story causes that it is believed more and more. In the case of Apollo, it is often embellished a little. What obviously has a very strong effect are the six following Moon missions, i.e. Apollo 12-17. I know many people for whom this bluff has massively increased the credibility, even if they all suffer from the same inconsistencies and even if the bluff and the anomalies in the pictures became bigger and bigger - towards the end the astronauts even drove around with a car that once left tracks and then didn't again. But the more NASA bluffed here, the less closely one looked.

NASA provides the continuation of its story just before the 40th anniversary with the lunar images of the LRO, as I described in chapter *7.2.2 Moon Images*. So the bluff is spun further, and despite the bad quality NASA reaps applause. By the way, the landing sites cannot be recognized as such from Earth even with the best telescope. A telescope would need a diameter of about 300 m to reach a resolution of 1 m over 380,000 km. This is also true for a telescope like Hubble, which is in Earth orbit. For a telescope in space this statement is exactly true, whereas a telescope with a diameter of 300 m on Earth would be disturbed by turbulences in the atmosphere, so that the achievable resolution would be reduced. Such optical calculations were also known in 1969, so NASA could be sure even then that its hoax would not be discovered anytime soon. Even if another nation with a Moon mission made photos from close and found nothing, then NASA could smilingly say, it had to photograph simply at the correct place, in order to see the traces of the Apollo landings.

Moon Landing, Conspiracy Theories and Wikipedia

NASA continues to produce stories reminiscent of the lunar landings even after the 40th anniversary. The LADEE mission, which sought to study the (virtually nonexistent) lunar atmosphere and the lunar dust within it, was justified by a drawing by Apollo 17 astronaut Eugene Cernan depicting dawn just before sunrise (Fig. 94). Every astronomer knows that there is no dawn at the Moon because of missing atmosphere; a star disappears in a matter of seconds when it is covered by the Moon and appears just as fast again.

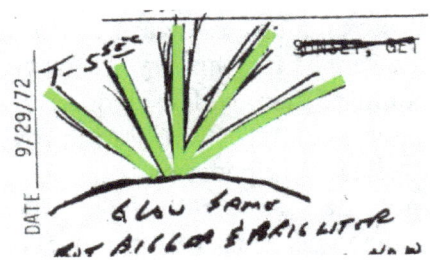

Fig. 94 Sketches of Sunrise during Apollo 17 [181]

The USA talk also again and again about manned Moon and Mars landings, in order not to let the topic dry up. As far as I can remember, these landings are always postponed for ten to fifteen years. Short enough to keep the budget going, but also far enough to avoid having to explain why the technology is not yet ready to carry out the promised landing.

"If the Moon landing was faked, I'm sure someone would have posted this on Wikipedia." I have heard this sentence many times. Instead of "Wikipedia", sometimes a newspaper or a journalist is brought into play, who would be only too happy to run such a story. But this is not the case, as I can confirm from my own experience. I have called the attention of various magazines and TV stations to the studio environment of the Apollo 11 pictures. Mostly my request was ignored. Only shortly before the 40th anniversary the Tages-Anzeiger, a newspaper published in Zurich, described first picture contradictions pointed out by me - after all a drop on the hot stone, which however is permanently kept hot by all the big newspapers and television by rehashing the official stories again and again. Surely a prominent insider, for example an Apollo astronaut, would have a better chance with the media. But the innermost circle or hard core seems to be really tough and has taken appropriate oaths. However, it is possible that less prominent insiders have also already informed the press, but that this information has not reached us.

But back to Wikipedia. This example shows how everything that does not fit into the mainstream concept is systematically deleted.

Apollo 11 – The Real Story

Wikipedia as a software platform is well done in itself. If you click on the tab "View history", you can see all previous versions and also compare them with each other. Further there is a tab "Talk", where users can make suggestions for improvement or ask questions. From my point of view Wikipedia has one big weakness: the authors are anonymous; most of them write under a pseudonym. It is not apparent whether someone is a private individual or part of an organization. This structure allows large organizations to control articles without anyone knowing who is behind the control. I have experienced censorship myself with the following example:

On December 15, 2013, under my username *Andisan,* I wrote a post in the German Wikipedia article *Conspiracy theories about the Moon landing* [182]. In doing so, I translated a section from a study by the University of Warwick into German, slightly reworded it, and added it to Wikipedia. Of course I have indicated which study it is and where it can be found. I was aware that you are not allowed to publish anything in Wikipedia that you have worked out yourself. One may only publish what comes from verifiable and reliable sources of information. [183] A university study seemed to me to be the most suitable, even if it was only published on the Internet and not in a scientific journal.

18 minutes later, my contribution had already been deleted with the argument "please provide proper evidence and leave your own opinions out of the article". The argumentation could not have been more inaccurate: my contribution summarized a part of a publication of the University of Warwick, thus did not describe my opinion at all, and I had given the reference to this publication. If one compares this, for example, with the preceding section, "Shadow casting", there only the view of the author is given, and there is no reference. The section starts as follows:

Many of the photos taken on the Moon show shadows cast by various objects that are not parallel to each other. Also, the length of the cast shadows is inconsistent. Conspiracy theorists interpret this as a contradiction to the sun being the only source of light on the Moon....

So, some conspiracy theorists are blamed for an interpretation that does not exist. Therefore, it is not possible to give a reference for this section. It remains, however, because it describes what we are supposed to read and believe here according to these watchers and their backers. Just at the example "shadow casting" the procedure of the Wikipedia truth twisters is to be recognized very nicely: they make vague hints to correct examinations of the Moon pictures, for instance such as I have described one in the chapter *4.4.3 Shadows from all Sides*. They change "shadows from all directions" to "non-parallel shadows", suggesting that "conspiracy theorists" would expect all shadows to be parallel in the photos, then they move the scenario to uneven surfaces, and finally they correctly explain the non-parallel shadows by perspective effects, so that the argument sounds plausible to the gullible and quick observer. I will give some more examples of truth twisting in Wikipedia.

I undid the deletion, as you can track, whereupon my post was deleted again by a third person using the same arguments. I knew that if I undid the deletion again, I risked being banned. But the example was worth it to me: I undid the deletion a second time and justified my action in the talk section. This time the article was deleted within two minutes and I was banned as an author. Again, you can see this by clicking on my username *Andisan*. My offense was that I had inserted unsubstantiated claims without giving sources, although I had given the source to the university study very well, as already described above. Several people were involved in this action on Wikipedia: two deleted my entries immediately, a third blocked me as an author, and others were involved in the talk.

Following this action, there was "coincidentally" a vandalism action, so that my post was no longer at the top of the version history and therefore not so easy to find. You can see this in the version history immediately after respectively above my posts from December 2013.

My reasoning is still available on the discussion or talk page, which can be clicked on in Wikipedia at the top left of each page [184].

Apparently, I had been a thorn in the side of the watchdogs for a while, since I had already replaced a link to a private website probably of one of these insiders in April 2013. This website had formerly had the Moon landing as a topic, but had then been converted and now advertised consumer goods. At that time I was still so careful to undo the deletion only once. So even after four days the link still pointed to the private website. Only when I duplicated then in the talk part, it was removed. This talk part can also be read in the Wikipedia Archive [185].

By the way, both talks were moved to the archive extremely quickly, so that they remained hidden from the normal user who does not browse the archive.

In summary, we can draw the following lessons:

1) The chaperones work in a coordinated manner and are always outnumbered.

Apollo 11 – The Real Story

2) Changes that deviate from the mainstream can only be found by systematically combing through the deleted versions and the comments, since what we are not supposed to know is not only deleted, but additionally covered up with further actions.

3) In many areas, you can find very good information on Wikipedia, but when it comes to explosive topics, Wikipedia only shows us the official representation, as we are also presented in the mainstream media.

These are three of the main lessons I have learned from the Moon landing story. However, the Moon landing is not an isolated case. I will come back to it in the next chapter.

By the way, Wikipedia was founded in 2001. It's a very good year to remember, because in 2001 an extremely drastic event took place in New York that had a major impact on global control and surveillance. Since September 11, 2001, we tolerate much more insight into our private data, and we allow ourselves to be controlled in more and more ways. When I once stood in an airport body scanner, had to hold my hands up in the air and was then patted down, I felt like a criminal who was being set up. We accept this because we are always told that this is necessary to fight terrorism.

I would now like to show you a few more examples on the basis of which we are led around by the nose. I continue to refer to (German) Wikipedia, specifically to the article "Conspiracy theories about the Moon landing". Wikipedia has the advantage that it is quite static and you can access it at any time - even earlier versions, as we have just seen. These examples can also be seen again and again in the mainstream media, i.e. in major newspapers and on television.

1) <u>Waving flag:</u> The procedure at Wikipedia is the same as with the shadow casting: again some conspiracy theorists are mentioned, who considered the movement of the flag as a forgery proof. And again there is no reference to a literature passage where such an accusation is made. I have also seen this argument in YouTube videos where a tarp or the flag moves as a result of the draft of passing astronauts. Here Wikipedia again takes only half, namely the movement of the tarpaulin or the flag and shows this at the example, where an astronaut handles at the flag and sticks the flag pole into the ground. That the flag moves in the process is more than logical, and that the flag is attached to a pole at its top and therefore appears nicely stretched is also logical. So if Wikipedia lists only half of the argumentation, even then the gullible reader thinks that

the "conspiracy theorists" must belong to a very simple-minded breed after all.
2) <u>Engine plume during launch from the Moon:</u> This argument also lacks a reference to a concrete accusation of a "conspiracy theorist". Actually, people from all walks of life expected an engine plume. NASA itself has drawn a clear engine plume at the emblem of Apollo 10 (Fig. 10). In other rocket launches and flights, a flame is permanently visible. The Wikipedia article shows as an example of an engine with little visible flame the main engine of the Space Shuttle, which burns oxygen and hydrogen to water, so it has a very clean burn. This flame is indeed less bright than the recoil cloud of the boosters, but still clearly visible [186]. According to Wikipedia, the ascent stage engine is also said to produce hardly any flame or other exhaust - whereas when the ascent stage is restarted, there is no flame visible at all and no exhaust fumes at all (Fig. 72 and Fig. 73).
3) <u>Invisible stars or starless images:</u> As with the previous two points, a generalized statement here is that the missing stars "have been interpreted as evidence that the footage was shot in a darkened hall." Again, there is no reference to a book or similar where someone has drawn this conclusion. Reading this, I don't know if this assessment is made up to portray "the conspiracy theorists" as stupid, or if it was made up by someone who has taken a serious look at the subject.

The question is not quite simple, because the stars are shown in a photo ideally as a very bright point: in a digital camera on a pixel and in the film used at that time on a grain. However, as a result of diffraction at the lens, in practice a star is imaged as a disk. In the center the disk is brightest and towards the outside the brightness decreases. One could now go and calculate the shape and size of the disk and then relate this to the graininess of the film. I did it differently. I photographed the sky at 1,000 meters above sea level on a clear night, using 200 ASA negative film. I first generously set the exposure to be right for a daytime shot (1/125 s, f/8). The sky was pitch black in the photo. Even when I increased exposure to 1s, f/1.7, the sky was still black. In other words, I don't expect to see any stars in the sky on the lunar images, where, after all, the exposure is set for daylight illumination, even though my film and camera were not identical to the Apollo 11 footage. So the reasoning on Wikipedia that no stars can be seen in the images is true.

Apollo 11 – The Real Story

However, the underlying claim could be fictitious as written above. Another question is whether an astronaut could see stars from the Moon. On this, Armstrong made the following statement at the post-flight press conference: „We were never able to see stars from the lunar surface or on the daylight side of the Moon by eye without looking through the optics." Aldrin doubled down, "I don't remember seeing any." To be sure, these statements are true for the two of them, since they were not on the Moon at all. However, for an astronaut standing on the lunar surface, they are false. When an astronaut on the Moon turns away from the Sun and looks up at the black sky, his eyes adjust to the dark background and he can then see the stars very well; almost as well as he sees them at night. But he must take a little time to do this. I myself have already seen the planet Venus in the blue sky at noon, when I happened to be looking at just the right place in the sky. It was a beautiful autumn day and I was about 1,000 m above sea level. So if I could spot Venus despite the blue sky background, it is easy for an astronaut on the Moon to spot stars on the black background of space.

On this subject, a former astronaut once mentioned at a lecture that he could have seen the stars from the ISS only at night. He mentioned this without anyone in the audience asking him. I had the impression that he would have had to say this on behalf of NASA so that Armstrong's statement at that time would be further supported. This also shows that today's NASA astronauts are in on the fake Moon landing. They know exactly what they must tell us, so that the old lies remain.

The Wikipedia article "Conspiracy theories about the Moon landing" mentioned here has been added to the list of excellent articles on December 23, 2006, as you can read at the bottom of the page. It was selected by Wikipedians, i.e. by the staff of Wikipedia, among others probably by those who had deleted my well referenced article in no time at all with the reason that it was not substantiated.

Another example of misinformation in Wikipedia is the question where the ascent stage of the Lunar Module has remained. According to the NASA information it remained in the Moon orbit, as I described and referenced in chapter 4.5. If one looks up on Wikipedia [187], it says "The ascent stage remained in lunar orbit and later crashed uncontrolled onto the Moon." Unfortunately, it does not say which ghostly hand caused the ascent stage to crash. A satellite orbiting the Moon does this for eternity, since there is no braking atmosphere at the Moon. However, Wikipedia has solved the problem with this statement that one would look for the ascent stage orbiting the Moon in vain.

In addition I must still note that the three-body problem is not solved until today. One cannot say with certainty that a satellite keeps its orbit for ever and ever, because in this case apart from the Moon also forces of attraction of the Earth, the Sun and the other planets act, and because also the Moon is not a perfect sphere. So it could happen that our Moon or a planet is thrown out of its orbit by the other celestial bodies. But in more recent times this has never happened; the orbits of the Moon and the planets have been drawing their circles, or rather their ellipses, constantly for several thousand years.

Here's one last point about Wikipedia: I was surprised to see Michael Collins' back surgery mentioned [98]. This alone is almost proof enough that he stayed on the ground. But I think it behaves here like with the artifacts in the Moon pictures. The details are publicly available. Depending on what you point your finger at, you can tell one version of the story or another. I was initially surprised that all the details were presented so transparently. But there is a passage in the relevant literature where exactly this effect has been described before. Hans Christian Andersen published the fairy tale "The Emperor's New Clothes" on April 7, 1837 [188]. Everyone could see that the emperor was naked, but no one wanted to admit it openly, because it would have made him look stupid. For only the intelligent are said to have been able to recognize the emperor's golden robes. It works the same way today: nobody wants to be counted among the species of stupid conspiracy theorists, so he prefers to switch off his own thinking apparatus and goes with the flow. Especially if he then looks up under Wikipedia, what conspiracy theory is exactly and what it contains. [189] Yes, he does not want to identify himself with it in any case and takes over with pleasure the version of the truth which is not provided with such a nasty attribute according to Wikipedia.

Another phenomenon of truth cover-up are the internet forums where the topic of Moon landing is allegedly discussed. Similar to Wikipedia, the authors are anonymous and act under a pseudonym. At the beginning of my research I put forward my arguments in such a forum, at that time still in the opinion that the participants would discuss this topic like me out of interest in the search for truth. Soon, however, I noticed that the fellows knew quite well where and to what extent they could engage in arguments and where they had to block. Many behaved - and still behave - as if they were being paid for their work. Permanently, a few are ready to give an answer to a newcomer. If there are no new requests, they start a new talk by themselves, so that the quick visitor thinks that the topic is really discussed. And as I said, they neatly avoid clear arguments, as I present them in this book.

Often the names of the watchdogs on Wikipedia are the same as those of the active writers on such Internet forums.

I observed a strong resemblance to Wikipedia in Amazon reviews. The German edition of this book was published on May 2, 2019. On May 16, Karl.M created an extremely competent 5-star review titled "New Evidence Against NASA Moon Landing". This review received many likes and was the first to be displayed on Amazon. This was a thorn in the side of the overseers, who apparently also exist at Amazon, and they created a confused 5-star review under the pseudonym "Hauptkommissar Freytag" (Chief Commissioner Freytag) with content that has nothing to do with my book. The title "Conspiracy theory or conspiracy practice?" and the subsequent text are anything but a recommendation for my book. Nevertheless, or precisely because of this, the overseers, who are organized in groups like on Wikipedia, always distributed so many likes that this review appeared and still appears as the first top review.

Since you can also order the reviews by "newest first", on 15 December 2022 an overseer under the pseudonym "Graf Zeppelin" (Count Zeppelin) wrote a scathing 1-star review, so that my book is now also poorly represented in this order.

So, the same applies to Amazon reviews as to Wikipedia: If you want to get to the truth, you have to go into depth.

I close this chapter with a quote from Upton Sinclair that reflects my experience in the space industry, where I worked for over 20 years, when I pointed out to colleagues that there were things wrong with the Apollo 11 Moon landing: "It is difficult to get a man to understand something, when his salary depends upon his not understanding it." [190]

8.3 Is the Moon Landing an Isolated Case?

As I already mentioned in the chapter 8.1 *General Conspiracies*, I consider the Apollo 11 Moon landing as a door opener to other topics, where worldwide the majority is lying. It was fascinating for me to see how clearly the facts are on the table, and how we nevertheless always manage to overlook them or to assess them as an exception or as a non-relevant individual case. In the following, I will list a few topics where lies are either also being told on a grand scale or which have the potential to do so. Behind these topics are interest groups that can earn money with it, expand their power or gain some other benefit from it. In general, I recommend always keeping an eye on both sides of such issues, that is, on the one hand, investigating the topic objectively and, on the other hand, asking the question "Cui bono?", i.e. "Who benefits?". It seems important to me to examine these two paths independently. I know many people who look at only one of these paths, for example, the question of benefit. If they don't find a quick answer, they conclude that there is no one who could benefit from it, and they don't even look at the facts and take the official view. With the Moon landing for example the benefit is not obvious. I have described it in chapter 7 from my today's point of view; this chapter is strongly personal colored. One cannot describe the benefit mathematically exactly - in contrast to a falsification of a picture. Everybody sees the benefit differently, even NASA itself made a neat overview of the scientific goals in 1969, and in 2018 their research chief casually says they just wanted to raise the US flag and leave footprints. So I recommend you again, look closely, even if you don't see a beneficiary at first glance. I like to compare such research with a criminal case. Imagine that the criminal police would first ask only for the beneficiary and only call in forensics when a beneficiary has been found. This procedure seems completely absurd; nevertheless, I observe this again and again with sensitive topics.

An indication that a topic is used for manipulation purposes is the label "conspiracy theory". With this extremely negatively occupied phrase it is achieved as already described earlier that nobody dares to deal with it openly in detail and therefore takes over the official version. However, as also said earlier, I ask you not to lump everything together. There are also conspiracy theories, which are intentionally obviously wrong, for example the Bielefeld conspiracy, which says that the city of Bielefeld does not exist at all, but that it is only faked. According to Wikipedia, this is a satirical conspiracy theory. [191] However, if one lists it together with the other conspiracy theories and chooses the heading "Do you believe in conspiracy theories", then it is gladly taken as a crowning conclusion to show how wrong *all* conspiracy theories are. An analogous example to the Bielefeld conspiracy is the topic *flat Earth*. Also this term seems at first sight only stupid, but it is used like the Bielefeld conspiracy to show other conspiracy theories as wrong. What I have seen with the *flat Earth* already is the following: first 100 arguments are enumerated, why the Earth is flat, and afterwards individual quite correct statements are made about other "conspiracy"-topics. The listeners should think then that after the first 100 wrong statements only further wrong statements can follow. Many people have already fallen for this clever method; one does not want to identify oneself with "flat Earth arguments".

Let's take a look at some other topics that are labeled "conspiracy theory" or that may not be questioned for other reasons. The list is neither conclusive nor complete. I show here topics that I have come across while doing research in the last 10 years:

<u>September 11, 2001 attacks in New York:</u> I have already mentioned these attacks a few times. They have a technical core similar to the Moon landing, so that if you look closely at the individual phases, you soon realize that a lot of lies were told. I mention only the most obvious example in my opinion, namely the collapse of the third building in Manhattan, the World Trade Center number 7 (WTC 7). No airplane ever flew into this skyscraper. It stood in the neighborhood of the two collapsed towers WTC 1 and WTC 2. According to the official version, a fire broke out in WTC 7 and then a pillar collapsed or something like that. However, WTC 7 collapsed as only a building collapses when it is blown up. You can watch the blast on YouTube. [192] You don't need to be a blasting expert to judge this. Common sense is enough - but that is exactly what they are trying to talk us out of by constantly calling in experts. The technical background of the attacks of September 11, 2001 is plausibly described in the book of D. A. Khalezov "9/11thology:..." in my opinion. [193]. As already mentioned in chapter 8.1 9/11 is in any case a conspiracy, because nobody has confessed to the attacks. However, one is considered a conspiracy theorist only if one considers a conspiracy in which the U.S. government or individual members of it were either privy to or actively involved.

<u>Man-made global warming</u>: The fact that it has become warmer in recent years is undisputed. Opinions differ as to whether the majority of the warming is caused by humans. There have always been warmer and

colder periods than today. By claiming that the majority of global warming is caused by pollutant emissions, it is possible to exert influence on the states and exercise control.

Kennedy assassination of 11/22/1963: President John F. Kennedy did at least two things that were contrary to possible background elites: he openly informed about conspiracies (see chapter 8.1) and he wanted to introduce interest-free money, which would have massively weakened the Federal Reserve. [194] Today's presidents no longer do this, in return they live longer.

UFOs: The in the chapter *7.3.3 Investment in Secret Projects* mentioned book UFOs [167] describes many UFO sightings. Together with other such books and reports this seems credible to me. UFOs, here flying objects with not conventional drives are meant, could come from visitors outside of the Earth, could be man-made or they could originate from a cooperation of extraterrestrials with humans. In the probably most well-known incident a UFO is supposed to have crashed in Roswell in the US state New Mexico in June or July 1947.[195] There is even a UFO museum there.[196] I have mentioned in the chapter 7.3.3 that Kennedy may have seen some UFO technology before he decided to announce publicly that we wanted to go to the Moon.

HAARP: The "High Frequency Active Auroral Research Program" consists of a huge transmitter (Fig. 95), with which the ionosphere, i.e. the uppermost part of the atmosphere, can be excited and set into oscillation. HAARP could have the potential to trigger natural disasters such as earthquakes, floods and volcanic eruptions. [197] The Fukushima tsunami may have been triggered by HAARP on March 11, 2011. Besides HAARP, explosions at the plate boundary between the Pacific Plate and the North American Plate could also have been used in a supporting role.

Fig. 95 Antenna Field of HAARP [197]

Condensation trails (chemtrails) and weather manipulation: Is the weather systematically influenced by additives in aviation fuel or is weather manipulation nowadays used covertly as a weapon? Hail missiles have been known for a long time, but what other technologies are there and what of them is used? [198] Weather weapons have been real for a long time, otherwise a UN resolution against such weapons would not have been passed in 1976. [199]

Health topics: These topics are now very much at stake for each of us, namely directly our own health and indirectly that of the animals and the plants, our food and nature in general.

Regarding topics on health and the health care system, the following questions arise:

1) Are the fertilizers and pesticides used still within the tolerable range or is, for example, glyphosate, one of the most widely used active ingredients in pesticides worldwide, sold under the brand name "Roundup", toxic and carcinogenic? [200]
2) Are all or at least almost all vaccinations good for the immune system and for people's well-being, or do they weaken people in the long term and make them all the more ill? [201] And what about, in particular, the many vaccinations given to babies in the first months of life? In any case, it would be more lucrative for the health industry's business model to add something to the vaccination so that the vaccinated person remains a customer in the future.

The most recent example are the so-called mRNA vaccines. Their toxicity is becoming more and more obvious. [228]

3) How quickly does mobile radiation damage our health? I have already seen some reports where calves were born blind, birds stopped breeding and bees died after a cell phone antenna was installed. In addition, there are more and more electrosensitive people, i.e. people who can no longer stand to be near a cell phone antenna. Among them are also people who had to deal with electrical devices at work and then suddenly developed health problems. The density and performance of mobile networks is constantly increasing. In 2018, we were at the fourth generation (4G or LTE for Long Term Evolution), and the fifth generation has been in the making since then. The reports of calf blindness, bee deaths and electro sensitivity have been around for several years. The bees were already dying at the low level of exposure at that time compared to today. Now the exposure is constantly being increased. How long will it take before the majority of the population is seriously harmed? If the transmitting power is apparently increased continuously without thorough

clarifications, this can be called a test experiment on living humans.

Surveillance: Especially since 2001, we have become more and more willing to allow the state and large corporations to view our personal data. Google now knows us better than our friends and, thanks to GPS tracking of our cell phones, always knows where we are. In Switzerland in the 1970s, a parliamentarian and army officer began collecting personal data and recording it on index cards known as fiches. [202] A political uprising followed. Today, most people reveal more about themselves via Facebook and similar platforms, let alone what data Google & Co collects via the Internet.

Fossil fuels: The adjective *fossil* is used to emphasize that it took a very long time for oil to form from organic matter. But oil is extracted from great depths and it is not clear whether and how organic material reaches such depths and how oil is formed from fossils. Oil could have been formed in another way [229]. But with the addition of *fossil,* the scarcity is emphasized and the population is willing to pay high prices. The oil industry is happy.

Migration: By flooding a country with migrants, one can damage this country, especially if these migrants are difficult to integrate, for example because they come from a different cultural background and/or do not speak the language. U.S. President Donald Trump, with his saying "America first," takes better care of his country than Angela Merkel, who would have invited migrants in (too) large numbers to come to Germany. I write "would have", because with me and probably many others the information arrived so in 2016, but I do not know how serious this information was. ZEIT-ONLINE wrote in an article at that time that it was different. [203] But however the migration flows have come about, the quality of life and social peace have suffered in many places. And there is also different information about this topic, especially about the living conditions in the countries of origin.

Money system: Hardly anyone understands how our money system really works. In Switzerland, similar to other countries, only the National Bank is allowed to issue money, but strictly speaking this means only money in the form of banknotes and coins. However, the share of cash in the money supply is small compared to the total money supply, of which, by the way, there are three: M1, M2 and M3. [204] Banks can conjure up money "out of thin air". [205] Although not in arbitrary quantities, banks play an essential role in the process of money creation. For example, if you take out a loan of 100,000 Euros with a bank, the bank writes "+100,000 Euros" in your account sheet. That's it. You now pay regular interest for this service, and at the end you have to return 100,000 Euros to the bank. In Switzerland, the monetary system was a topic in June 2018: the popular initiative "Vollgeld" (sovereign money) was voted on. The initiative was rejected, but the positive effect was that people thought about the issue.

When lending money to private individuals, the above procedure may make sense. But with semi-private central banks like the Federal Reserve (FED) [206], which sucks its economy dry using the same principle, namely the interest it charges, you don't know what it uses its profits for. This is particularly the case in times of high interest rates.

With all the topics listed, many questions remain unanswered. Even if it is unclear who the beneficiaries are and what exactly they gain from it, these are all topics that have a high potential for disinformation and manipulation. Many serve to make people more controllable; or weaken us. Take a closer look and inform yourself.

For example, in the September 11, 2001 attacks, the NATO alliance case was triggered because New York was attacked; New York is in NATO, so NATO was attacked, and consequently NATO member countries had to help defend NATO. Therefore, Germany went to war in Afghanistan, where Bin Laden is said to have been hiding. As a German, I would be furious if "my" country went to war on the basis of such shaky information, and I would find it doubly tragic if, for example, an acquaintance of mine had died there.

When it comes to the healthcare system, one might wonder whether the health of the population is really the top priority. Imagine that you had just found a very inexpensive cure for cancer. Do you think the healthcare industry would be happy about your discovery? The cost of illness would fall, and with it the gross national product. And it is the gross national product that usually carries the most weight in the ranking.

8.4 Has there ever been a Man on the Moon?

The analyses of the NASA Apollo documentation have shown that all parts of the described missions, which took place beyond the Earth orbit, are fictitious. Really real is only the facade, which in the case of Apollo consisted of immense activities on Earth and Saturn V rocket launches. The Apollo astronauts behaved as actors. It's fitting that a star is dedicated to the Apollo 11 crew on the Hollywood Walk of Fame. [221] Rocket technology in the 1960s was not yet so advanced that it would have been possible to launch back from the Moon and dock with an orbiter within a day at most. Nevertheless I ask here the question whether already once a man was on the Moon. Perhaps there was and is indeed a superior technology that makes this possible. I have described in the chapter *7.3.3 Investment in Secret Projects* that Kennedy, according to the book *...The Lion Sleeps No More,* is supposed to have seen this technology shortly after his election. [166] I have read similar things in other books. They range from UFOs as flying saucers with antigravity propulsion or field propulsion to wormholes and teleportation. For me it still sounds very speculative to assume that people on Earth master such technologies and use them in secret. UFOs of extraterrestrial origin, as Erich von Däniken describes them in his books, seem more probable to me. Already in the Indian mythology flying machines were described, so-called Vimanas. In the history lessons in the school I have never heard of it. There everything was and still is omitted, which does not fit into the world view of a linear development from the Neanderthal man to the today's technologized man. But maybe there was a high culture on the Earth x-thousand years ago, which mastered such technologies; and maybe these people were already once on the Moon.

I can neither confirm nor exclude that there were or are such hidden technologies.

If we already deal here with the question whether ever men were on the Moon, a follow-up question would be whether there are any signs on the Moon which point to something artificial or unnatural. Here the so-called crater rays come into my mind. One recognizes these on many Moon pictures, best with full Moon (Fig. 96). I have found a study about this [207]. At the beginning one assumed that these rays originate from material which was ejected during a volcanic eruption. Large chunks would have created secondary craters, which in turn would have produced ejected material (in the same direction). Alternatively, these craters could have been formed along tectonic fractures. Now, however, one is of the opinion that the ejecta material was ejected by impacts of meteorites. The craters consist of young rock and are therefore bright. Likewise the ejected material is young and bright and was deposited on the Moon surface as a thin layer. These attempts of explanation do not satisfy me. For me it is still an open question why then such thousand-kilometer-long dead-straight lines have developed.

Apollo 11 – The Real Story

Fig. 96 Moon with Crater Rays

We never see the back side of the Moon from the Earth. The Moon always shows us the same face. It rotates around its own axis exactly once a month, because its rotation is bound to its orbit. Since its orbit is elliptical and inclined, we see a little more than half, 59% to be exact, of its surface over time. [208]

I searched for pictures of the back side of the Moon and found the following picture (Fig. 97), which came from Apollo 16:

Fig. 97 would have to be composed of different images from Apollo 16, because Apollo 16 was never so far behind the Moon to take such a single image. Interestingly, no crater rays can be seen here.

Which probe delivered the data for the Apollo 16 image, I don't know; it was probably not Apollo 16, but rather another (unmanned) probe.

No crater rays are visible in a similarly composite image of the LRO (Fig. 98). So the question is, why are the crater rays only on the side facing the Earth?

Fig. 97 Back Side of the Moon photographed by Apollo 16 [209]

Fig. 98 Back Side of the Moon photographed by the LRO [209]

Traces of people, buildings or similar are not visible on the Moon images. Such traces would have to be large. The highest available resolution of the LRO image is 18,000 x 18,000 pixels. With a lunar diameter of 3'476 km, one pixel then has an edge length of 193 m. Everything smaller than about 200 m disappears in one pixel and is no longer recognizable.

8.5 Conclusion

The Apollo Moon landings are *the* poster story in the major Western media for an assured historical event. Colleges and universities also repeatedly present these heroic stories as fact. These stories are even presented as so certainly true that one may not even question them without being put in a bad light. The situation is similar in science, where once-accepted viewpoints are to be blindly accepted [210]. I once asked at an aperitif a university professor who had examined Moon rocks whether he could have distinguished Moon rocks from terrestrial rocks. "I don't discuss the authenticity of the Moon landing", was his reply. He turned on his heel and walked away without saying goodbye.

As an expert in the field of space travel, I dared to check the NASA documentation for its authenticity, i.e. I asked myself whether the documentation was authentic or whether it contained contradictions. The result is more than clear. The story presented by NASA is teeming with contradictions, where actually there should not be a single gross contradiction. The used technology is presented by NASA in a maturity which could not be reached by NASA itself neither before nor in the following 40 to 50 years. I mention here again the most striking example, namely the time from the launch of a spacecraft to a successful rendezvous. This time was a little more than two days until 2013 and could then be reduced to 6 hours for the special case that the target satellite is just passing the launch site during the launch. However, for the Moon this time is said to have been only 3.41 hours and for Gemini 11 even only 1.34 hours. The time of Gemini 11 is 30 times shorter than that achieved today, because the Agena target satellite flew sideways past the launch site during the launch of Gemini 11. The discrepancy of a factor of 30 is like when NASA claimed that its most athletic employee jumped 267 m in the long jump in 1969; Bob Beamon managed 8.90 m in 1968. The logical consequence of inability to fly to the Moon with the technology presented is that the images of the lunar surface come from a studio and the experiments carried out there are the product of NASA scientists' imagination.

I am not the first or the only one who has come to this conclusion. However, my predecessors could hardly prevail with their results so far. After all, according to WikiSpooks, a Wikipedia-like platform, the teaching opinion is not the same in all countries. For example, wherever Cuban teachers teach, they teach that the Moon landing was faked. [211] On the same page there are also survey results that give a similar picture as the one I quoted in the introduction.

Moon Landing, Conspiracy Theories and Wikipedia

So I wrote another book on this subject according to the motto "Constant dripping wears the stone". I followed the NASA documentation and kept the description as transparent and comprehensible as possible. This was not possible in former times without a well-functioning Internet. I hope you have thought along critically while reading this book and checked whether I have proceeded correctly.

In the introduction, I distinguished three cases regarding the truthfulness of the Apollo documentation: transparent success, embellished success and staging. The third variant "staging" fits perfectly to the Apollo 11 mission. The second variant "Embellished success" implies that at least one manned Moon landing including return to Earth had taken place under exclusion of the public. I cannot exclude this, but such a mission would have had to be carried out with another technology than the known rocket technology.

If I now state with this important topic that we are lied to by NASA and the news agencies on edge, then I have probably rather stumbled on the tip of a (global) lie iceberg than that I have uncovered the worldwide only lie structure. In my research activity I found many sources which mentioned other topics besides the Moon landing. I have addressed these in the chapter 8.3 *Is the Moon Landing an Isolated Case?*.

If I was able to show you the tip of an iceberg in this book, of which you perhaps did not even know with certainty that it exists at all, then the question arises as to what the situation is with the 90% of the iceberg that is under the surface of the water. The knowledge about actually taking place disinformation and manipulation can open you now the entrance to respectable sources which you would have classified so far as suspicious; whereby you are not spared of course to find out yourself whether a source reports respectably.

Serious reporting refers to specific cases, provides sources and substantiates its arguments. Dishonest reports are often general, do not give sources, compare topics that have nothing to do with each other factually and, for example, are only related to each other by the label conspiracy theory. Often unserious sources make assertions out of thin air, which are then also correctly substantiated.

I have often told someone that WTC 7 collapsed on September 11, 2001 in a way that was only possible with a demolition. To this I received as an answer: "And do you also not believe in the Moon landing?". You see, conditioning has worked. The word "believe" here is not only etymologically related to religious faith. We have been brought up to dutifully tell the truth. Therefore, many cannot or do not want to imagine covert programs.

That there was at least one such covert program we have seen in this book. If by now you have watched the video of the collapse of WTC 7 [192] and agree with me that it was blown up, then we already have two events in which completely or partially wrong information was given; by the way, just in the two events in our generation, which everybody can remember exactly and also still knows where he was at that time. So there are one or more organizations on this Earth, which are able to lead us around by the nose with such events. These organizations, I choose for the sake of simplicity only the plural, are not only able to make us believe false events, they also control directly or indirectly the news agencies, press and also governments. Some authors speak of so-called background elites. These two events are obvious proof that they exist. Either they are the heads of these organizations or, if the hierarchy does not stop there yet, these background elites are above these organizations.

The next question that now arises is, what is the purpose of these background elites? I can't answer this question conclusively either, but I can give you some hints about it, from which one can deduce where it might be going. Certainly, like any ruler, these elites want to consolidate and, if possible, expand their power. They operate in the background and act against the welfare of mankind.

Let's assume that these elites controlled not only the Moon landing and at least parts of the September 11, 2001 attack in New York, but additionally events and issues that I discussed in chapter 8.3 *Is the Moon Landing an Isolated Case?*. If they wanted to further consolidate and expand their power, they would have to make sure that, if possible, the whole world fits into an organizational chart, and that all countries and large organizations are combined into a hierarchy. Then these elites could give orders from above, which would then be carried out below, as is already the case today in armies, in states and in companies. The goal would have to be that, unlike today, countries would no longer be independent, but would have to subordinate themselves to this super-organization. Isn't this scenario exactly what we are currently observing on the world stage? Weren't countries that didn't want to subordinate themselves, such as Libya, destroyed?

Moon Landing, Conspiracy Theories and Wikipedia

Many events in this world can be better classified if you look at them from the point of view that the unspoken goal is global control. If, for example, cash is to be abolished and people can then only pay electronically, then the flow of money can be perfectly controlled, and an operator can "at the push of a button" block people who are not agreeable to the system.

So what else I have done at the end of this book, that is, after analyzing the Apollo 11 mission, is once again exactly the opposite of what the background elites pretend. Instead of sticking to the unspoken addendum "the official story is the truth; questioning forbidden" when it comes to topics labeled "conspiracy theory", I have taken up exactly some of these topics and dared to ask questions and scratch the surface here as well. To go deeper would have been beyond the scope of this book. However, if I have succeeded in making you aware of these issues in addition to space travel, then writing this book has been doubly worthwhile. Have the courage to use your own mind, even if a topic is labeled conspiracy theory. The knowledge imparted here puts you above that ridiculous labeling.

I close this book in the spirit of John F. Kennedy with the confidence that with your help man will be what he was born to be: free and independent.

9   About Me

I was born in 1955 and grew up in a suburb of Zurich (Switzerland) as the youngest of three siblings. After elementary school, I attended high school in Zurich and then studied electrical engineering at the Swiss Federal Institute of Technology. After graduation, I took a job with a company developing anti-aircraft systems. The same company also established an aerospace department, to which I transferred in 1995. There I worked on the development of instruments and devices that would later be bolted onto satellites so that they could also measure, transmit data and take pictures. In 2018, I left the space company to work as an independent author.

I am married and the father of two grown-up daughters. After a colleague drew my attention in 2008 to possibly recreated images of the Apollo missions, I decided to get to the bottom of this statement. In the process, I worked my way deeper and deeper into the subject of manned spaceflight and the experiments it made possible, such as "Lunar Laser Ranging". Ten years later, I had so much material together that I decided to write a book.

10 Appendix

10.1 Lunar Orbit

The Moon orbits the Earth at an average distance of 384,000 km. This corresponds to 110 lunar diameters. Thus, we see the full Moon at an angle of half a degree. The Moon's orbit is slightly elliptical, so the distance varies by ± 20,000 km. The Moon's orbit is inclined at 5° with respect to the ecliptic (plane of the Earth's orbit around the Sun). However, the axis of rotation of the Moon is inclined only 1.5° with respect to the ecliptic; therefore, the Moon knows no seasons. In comparison, the Earth's axis is inclined to the ecliptic by 23.5°. [212][213]

When the Apollo spacecraft arrived at the Moon on July 19, 1969 17:22 Universal Time, it was 1° below the ecliptic and when it left lunar orbit on July 22 04:58 it was 3° below. The distance of the Moon was on July 19 391,000 km and on July 22 388,000 km [214], slightly above the mean value of 384,000 km.

10.2 Data of the Earth and the Moon

In Tab. 1 is a collection of data about the Earth and the Moon which is always needed in connection with a flight to the Moon.
The parameters labeled "➔" are derived from the top three values.

Apollo 11 – The Real Story

Tab. 1 Data of the Earth and the Moon

	Earth	Moon
Radius	6,370 km	1,738 km
Mass	$5.976 \cdot 10^{24}$ kg	$0.0123 \cdot$ Earth mass
Gravitational constant Γ	$6.674 \cdot 10^{-11}$ m^3 / (kg $\cdot$ s^2)	
➔ Acceleration on the surface	9.82 m/s^2	1.62 m/s^2
➔ Velocity [a)] and orbital period of a satellite with a circular orbit and an orbital altitude of • 100 km • 300 km • 1000 km	7.85 km/s - 86 min 7.73 km/s - 90 min 7.36 km/s - 105 min	1.63 km/s - 118 min 1.55 km/s - 138 min 1.34 km/s - 214 min
➔ Time it takes for a stone to touch the ground when dropped from 1m height	0.45 s	1.11 s (2.5 times longer) [b)]
➔ Jump height when jumping with an initial speed of 5 m/s	1.3 m	7.7 m (6 times higher)
➔ Horizontal visibility on a sphere of the same size, e.g. at the sea, from a height of 1.5 m, to the horizon	4.4 km	2.3 km

a) Often you see data in km/h. The conversion is as follows:
7.85 km/s $\cdot$ 3600 s/h = 28,260 km/h

b) If a film recorded on Earth is played back 2.5 times slower, flying bodies will get the impression that the film was recorded on the Moon.

I used the following formulas to determine the derived quantities:

Appendix

Tab. 2 Formulas used in Tab. 1

Acceleration on the surface of the Earth	$g = \dfrac{\Gamma \cdot Mass\_Earth}{Radius\_Earth^2}$
Speed of an Earth satellite on a circular orbit	$v = \sqrt{\dfrac{\Gamma \cdot Mass\_Earth}{(Radius\_Earth + Orbit\_Height)}}$
Associated circulation time	$T = 2 \cdot \pi \cdot \sqrt{\dfrac{(Radius\_Earth + Orbit\_Height)^3}{\Gamma \cdot Mass\_Earth}}$
Time to fall to the ground from the height h	$T = \sqrt{\dfrac{2 \cdot h}{g}}$
Jump height with initial velocity v_0	$h = \dfrac{v_0^2}{2 \cdot g}$

The visibility is calculated as follows:

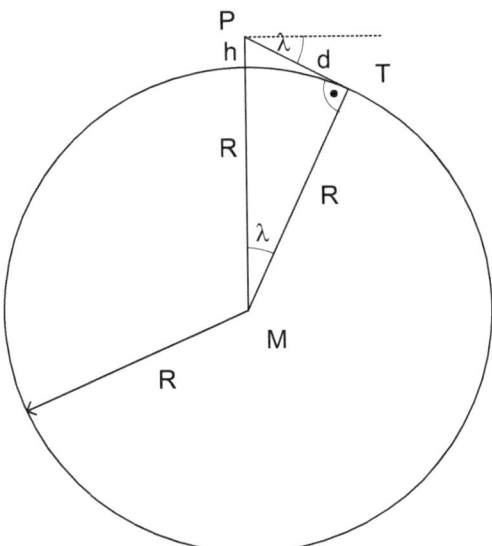

Fig. 99 Determination of the Horizontal Visibility on a Sphere

Apollo 11 – The Real Story

Visibility from the height h
(from observer P to T, the farthest visible point on the sphere):
$$R^2 + d^2 = (R+h)^2$$
$$\Rightarrow d = \sqrt{2 \cdot R \cdot h + h^2} \approx \sqrt{2 \cdot R \cdot h}$$
Example Earth, h=1.5m: $d \approx \sqrt{2 \cdot 6370km \cdot 0.0015km} \approx 4.4km$
Example Moon, h=1.5m: $d \approx \sqrt{2 \cdot 1738km \cdot 0.0015km} \approx 2.3km$

Visibility at unknown height h, but given angle λ (with respect to the horizontal):
1. calculation of h:
$$R + h = \frac{R}{\cos(\lambda)}$$
$$\Rightarrow h = \frac{R}{\cos(\lambda)} - R = R \cdot (\frac{1}{\cos(\lambda)} - 1)$$
2. insert into above formula (d≈ ...):
$$d \approx \sqrt{2 \cdot R \cdot R \cdot (\frac{1}{\cos(\lambda)} - 1)} = R \cdot \sqrt{2 \cdot (\frac{1}{\cos(\lambda)} - 1)}$$

Example Moon:

$$\lambda = arctg\left(\frac{1.5m}{34m}\right) = 2.5° = 44 mrad \Rightarrow h = 1600m, d = 76km$$

(or calculate the arc directly: $\lambda \cdot R$ = 0.044 rad · 1738 km = 76 km)

Appendix

10.3 Dimensions of the Lunar Landing Module (LM)

The height is 7 m (22 feet 11 inches) with the supports extended. The distance between two diagonally opposite feet is 9.4 m. The descent stage has a height of 3.2 m and a diameter of 4.2 m. For the ascent stage, the height is 3.8 m and the diameter is 4.3 m. [38]

10.4 Separation Maneuver

Fig. 100 shows the diagram mentioned in chapter 4.3 which shows the relative motion of the two spacecrafts. At the top it says "LM ABOVE AND BEHIND" which is equivalent to "CSM below and in front". The latter fits better, since it is the CSM that is drifting away from the Lunar Module LM.

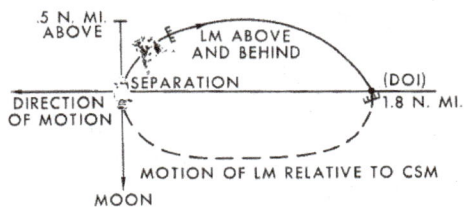

Fig. 100 Movement of the Lunar Landing Module (LM) as seen from the CSM [28]

At the beginning, the CSM drifts towards the Moon and only with time also towards the front.

The motion continues to the DOI (Descent Orbit Insertion) point where the Lunar Module enters descent orbit. The dashed line would only have come into play if the Lunar Module had not been able to descend and had continued in free flight.

10.5 Test Resolution to Fig. 54

I faked the sunrise in Fig. 54. The picture with the foggy background is real, but I copied the Sun into it. The Sun doesn't rise that low down there.

Fig. 101 shows what it looks like at this location in good weather:

Fig. 101 Landscape as for Fig. 54, but with nice Weather

10.6 Confirmation that the Wide-Angle Lens was used for Live Video

The confirmation that indeed the wide-angle lens was used for the live transmission is provided by the TV image itself: the distance from the TV camera to Aldrin is about 3 m, as I have estimated from the training images in Fig. 59 and Fig. 60. For this I also took the distance of the two feet of the Lunar Landing Module, which is 6.6 m, to help. If you imagine a 30 cm long ruler in front of Aldrin's chest, this ruler covers about the width of the chest under the suit. This ruler would be seen under an angle of 100 mrad or 5.7°. If one measures the image diagonal with it, one gets quite exactly the indicated 80°. The objective with the next smaller field of view, the "Lunar Day" objective, has a diagonal of 35° [69], which is more than two times smaller. With this lens, the TV camera would have had to be about 7 m away from Aldrin to image him at this size, which no longer fits this situation. This estimation thus clearly confirms that the wide-angle lens was indeed used for the live broadcast, as recommended in the camera manual.

10.7 Easier Rendezvous Option for Gemini 11?

In this chapter I pursue the question whether it would have been possible that the Agena rocket could have started in such a way that it would have flown after an Earth orbit or a little bit more straight again over Cape Canaveral, thus over its starting place. With this constellation Gemini 11 could have ascended directly into the same orbital plane, could have saved itself the change of the orbital plane and could have reached the rendezvous faster and with less fuel.

Indeed, one can imagine such an orbit. If the Agena rocket as shown on Fig. 102 had been launched slightly to the north (launch 1), it would have flown over Cape Canaveral again (launch 2) after a little more than one orbit, and from there Gemini 11 could have launched slightly to the south. Everything would have taken place in one orbital plane. The inclination of this orbit is 0.6° larger, so 29.6°.

NASA, however, has clearly described in chapter 4.5 with the Fig. 69 and Fig. 70 the variant with orbits having an inclination of 29°. Such orbits result when a satellite launches due east. For Gemini 8 as well as for Gemini 11 the inclination is given with several digits after the decimal point; mostly just under 29°, which corresponds exactly to the latitude of Cape Canaveral.

Appendix

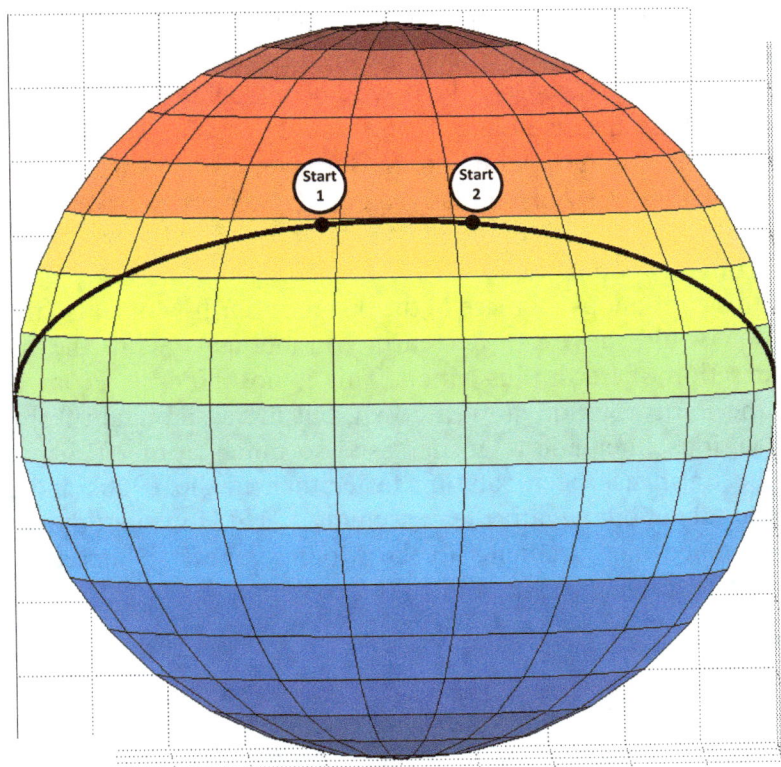

Fig. 102 Possibility of a Common Orbit

I think, one cannot "save" the mission nevertheless, i.e. think, NASA would have flown this variant with the inclination 29.6° in secret and so achieved that the space capsules could dock as reported in very short time. Even with the optimal initial constellation, where the trajectory of the target satellite passes straight over the launch point, it took a little more than two days to dock until 2013, so even then the reported times would be beyond what was possible in the 1960s.

The development of technology in the 50 years after Apollo 11 also speaks a clear language on this point.

10.8 Should the Moon move away from the Earth?

Physically it is not at all plausible that the Moon should move away from the Earth with a more or less constant speed. What we can observe are the tides. The sea level and also the Earth's crust rise and fall as a result of the Earth's rotation and the Moon's orbit. This creates frictional heat at the expense of kinetic energy. If one considers the system Earth plus Moon isolated, thus neglects the influences of the other planets and the Sun, then this frictional heat would indeed slow down the Earth rotation and the distance Earth-Moon would increase. This follows from the principle of angular momentum. But I consider this isolated view as inadmissible, since only already the influence of the Sun is clearly visible: at full respectively new Moon the tides are more pronounced than at other Moon phases.

With a continuous increase of the distance Earth-Moon also the Earth rotation would have to decrease continuously with the isolated consideration of Earth plus Moon. This is not the case. From 1965 to 1972 the Earth rotation slowed down, but increased again from 1978. After various fluctuations, it increased so much from 2016 that since then the Earth has been rotating faster than in the 1960s and people began to talk about negative leap seconds. [216] However, the invented laser distance measurements to the Moon suggest an increase of the Moon orbit at the expense of the day length.

10.9 Notes to this English Edition

In 2018, I wrote a book about the coming 50th anniversary of the Apollo 11 mission in German, my mother tongue. In order to make this book accessible to the English-speaking population, I translated it into English. The result is the book which you are reading now.

I took the opportunity to make small additions here and there. But this book still remains to 98% a translation of my initial German book 50 JAHRE APOLLO 11 MOND-(F)LÜGE what could be translated as 50 YEARS OF APOLLO 11 MOON (F)LIES.

11 References

I worked with these references in the period April to November 2018. Some hyperlinks disappeared during this time. In this case, I refer to the last version in the web archive (http://web.archive.org). I checked the hyperlinks for the last time on 11/11/2018.
You can download these references under the name "Apollo11References.pdf" on my homepage https://apollophotos.ch/en, so you can open them simply by clicking on them.
Wikipedia references are taken from the German Wikipedia if nothing else is mentioned. During the translation into English, I added references 215 ff from September 2022.

[1] NASA: Explorer-I and Jupiter-C (https://history.nasa.gov/sputnik/expinfo.html)

[2] August 22, 1963 (https://www.nasa.gov/centers/dryden/history/thisweek/EC68-1937.html)

[3] E-5251 (https://www.nasa.gov/centers/dryden/multimedia/imagegallery/X-15/E-5251.html)

[4] Neil Armstrong with X-15 #1 After Flight (https://www.nasa.gov/centers/dryden/multimedia/imagegallery/X-15/E-USAF-Armstrong-X-15.html)

[5] Wikipedia: Woschod 2

[6] NASA: A History of Spacecraft Environmental Control and Life Support Systems (https://ntrs.nasa.gov/archive/nasa/casi.ntrs.nasa.gov/20080031131.pdf)

[7] NASA: Gemini IV (https://www.nasa.gov/multimedia/imagegallery/image_feature_1061.html)

[8] NASA: Gemini IV (http://web.archive.org/web/20130306104754/https://nssdc.gsfc.nasa.gov/nmc/masterCatalog.do?sc=1965-043A)

[9] Ed White: First American Spacewalker (https://www.nasa.gov/multimedia/imagegallery/image_feature_838.html)

Apollo 11 – The Real Story

[10] Gemini 8 (https://history.nasa.gov/alsj/alsj-GeminiVIII.html)

[11] Gemini 8 Mission Report (https://history.nasa.gov/alsj/43455667-Gemini-Program-Mission-Report-Gemini-Viii.pdf)

[12] NASA: Lunar Orbiter 1 (http://web.archive.org/web/20150923071513/https://nssdc.gsfc.nasa.gov/nmc/masterCatalog.do?sc=1966-073A)

[13] NASA: Gemini 11 (http://web.archive.org/web/20111022062300/https://nssdc.gsfc.nasa.gov/nmc/masterCatalog.do?sc=1966-081A)

[14] Magazine 48/X (B & W) (https://www.hq.nasa.gov/alsj/a12/images12.html#Mag48)

[15] history of europe in space (https://www.esa.int/About_Us/Welcome_to_ESA/ESA_history/History_ESRO-1_satellite_1968)

[16] Wikipedia: European Space Research Organisation

[17] John F. Kennedy "Landing a man on the Moon" Address to Congress - May 25, 1961 (https://www.youtube.com/watch?v=TUXuV7XbZvU)

[18] NASA: The Apollo Missions (https://www.nasa.gov/mission_pages/apollo/missions/index.html)

[19] NASA: Apollo 1 (https://www.nasa.gov/mission_pages/apollo/missions/apollo1.html)

[20] NASA: Apollo-Saturn Uncrewed Missions (https://www.nasa.gov/mission_pages/apollo/missions/Apollo-Saturn-Uncrewed.html)

[21] Apollo 11 Mission Report (https://ia801209.us.archive.org/7/items/NASA_NTRS_Archive_19710015566/NASA_NTRS_Archive_19710015566.pdf)

[22] Apollo 204 Accident Report (https://history.nasa.gov/as204_senate_956.pdf)

[23] NASA Armstrong Recalls First Moon Landing, Preps For 'Next Giant Leap' (https://www.nasa.gov/centers/armstrong/Features/armstrong_recalls_first_moon_landing.html)

References

[24] Landing Safely After Rehearsal Mishap (https://www.nasa.gov/centers/johnson/multimedia/aod/S68-31666.html)

[25] The Flight Of Apollo 7 - NASA Documentary (https://www.youtube.com/watch?v=lxK4BAbDQf0)

[26] Saturn V (https://www.nasa.gov/centers/johnson/rocketpark/saturn_v.html)

[27] Saturn V Second Stage (including a diagram of Saturn V) (https://www.nasa.gov/centers/johnson/rocketpark/saturn_v_second_stage.html)

[28] Press Kit; Apollo 11 Lunar Landing Mission (https://www.hq.nasa.gov/alsj/a11/A11_PressKit.pdf)

[29] The F-1 Engine … (https://www.nasa.gov/topics/history/features/f1_engine.html)

[30] LAUNCH VEHICLES (https://www.hq.nasa.gov/alsj/CSM02_Saturn_Launch_Vehicles_pp8-14.pdf)

[31] NASA: LAUNCH ESCAPE (https://www.hq.nasa.gov/alsj/CSM15_Launch_Escape_Subsystem_pp137-146.pdf)

[32] Command Module (https://www.hq.nasa.gov/alsj/CSM06_Command_Module_Overview_pp39-52.pdf)

[33] Reaction Control System (https://history.nasa.gov/afj/aoh/aoh-v1-2-05-rcs.pdf)

[34] Apollo 11 Command and Service Module (CSM) (http://web.archive.org/web/20170219055508/https://nssdc.gsfc.nasa.gov/nmc/spacecraftDisplay.do?id=1969-059A)

[35] Apollo CSM (https://web.archive.org/web/20071217222825/http://www.astronautix.com/craft/apolocsm.htm)

[36] Using History to Design the Future (https://www.nasa.gov/mission_pages/constellation/orion/umbilical_inspection.html)

[37] Apollo 11 Lunar Module / EASEP (http://web.archive.org/web/20170223210708/http://nssdc.gsfc.nasa.gov/nmc/spacecraftDisplay.do?id=1969-059C)

[38] LUNAR MODULE (https://www.hq.nasa.gov/alsj/CSM08_LM_&_SLA_Overview_pp61-68.pdf)

[39] APOLLO EXPERIENCE REPORT -- ASCENT PROPULSION SYSTEM (https://ntrs.nasa.gov/archive/nasa/casi.ntrs.nasa.gov/19730010173.pdf)

[40] Apollo Lunar Module Propulsion Systems Overview (https://ntrs.nasa.gov/archive/nasa/casi.ntrs.nasa.gov/20090016298.pdf)

[41] Saturn V First Stage (https://www.nasa.gov/centers/johnson/rocketpark/saturn_v_first_stage.html)

[42] SATURN V LAUNCH VEHICLE FLIGHT EVALUATION REPORT-AS-506 APOLLO 11 (http://web.archive.org/web/20180504141219/https://www.ibiblio.org/apollo/Documents/lvfea-AS506-Apollo11.pdf)

[43] Apollo 11 Mission Overview (https://www.nasa.gov/mission_pages/apollo/missions/apollo11.html)

[44] https://sonnen-sturm.info/hilfe_faq/glossar

[45] BIOMEDICAL RESULTS OF APOLLO (https://history.nasa.gov/SP-368/contents.htm)
→ Chapter 3: RADIATION PROTECTION AND INSTRUMENTATION (https://history.nasa.gov/SP-368/s2ch3.htm)

[46] http://z-e-i-t-e-n-w-e-n-d-e.blogspot.ch/2013/10/24-jahre-nach-dem-quebec-blackout-durch.html

[47] Super Sonnensturm – Ein Rückblick auf das Carrington-Event von 1859 (https://sonnen-sturm.info/super-sonnensturm-ein-rueckblick-auf-das-carrington-event-von-1859)

[48] Cesium Iodide Dosimeters (http://web.archive.org/web/20080923051140/https://nssdc.gsfc.nasa.gov/nmc/experimentDisplay.do?id=1966-073A-04)

References

[49] Radiation Analysis for the Human Lunar Return Mission
(https://ntrs.nasa.gov/archive/nasa/casi.ntrs.nasa.gov/19970031679.pdf)

[50] Radiation Analysis for Moon and Mars Missions
(https://arxiv.org/abs/1805.01643)

[51] Astronaut Still Photography During Apollo
(https://www.history.nasa.gov/apollo_photo.html)

[52] APOLLO-11 HASSELBLAD CAMERAS
(https://history.nasa.gov/alsj/a11/a11-hass.html)

[53] Apollo 11 Image Library
(https://www.history.nasa.gov/alsj/a11/images11.html)

[54] Apollo 11 Photography Index
(https://history.nasa.gov/afj/ap11fj/photos/37-r.html)

[55] Datenblatt "Planar T* f/2.8- 80 mm"
(http://www.hasselbladhistorical.eu/pdf/lds/CF80.pdf)

[56] Descent Flight Path
(https://www.hq.nasa.gov/alsj/a11/a11_descent.jpg)

[57] Apollo Landing Site Coordinates
(https://nssdc.gsfc.nasa.gov/planetary/lunar/lunar_sites.html)

[58] Bezos gelingt Testflug mit wiederverwendbarer Rakete
(http://www.spiegel.de/wissenschaft/weltall/blue-origin-gelingt-test-mit-wiederverwendbarer-rakete-a-1064447.html)

[59] Bitte einmal Weltraum retour!
(https://www.nzz.ch/panorama/spacex-bringt-rakete-kurz-nach-start-erfolgreich-zur-erde-zurueck-1.18666992?extcid=Newsletter_22122015_Top-News_am_Morgen)

[60] Historic Rocket Landing
(https://www.youtube.com/watch?v=9pillaOxGCo)

[61] SpaceX
(https://twitter.com/spacex/status/943945832063045632)

[62] Wikipedia: Liste der künstlichen Objekte auf dem Mond

[63] Apollo 11 Video Library
(https://www.hq.nasa.gov/alsj/a11/video11.html) Das Video mit der besten Qualität zeigt Aldrin beim Aussteigen und ist bei 109:42:28

[64] Apollo 11 Lunar Surface Journal (https://www.hq.nasa.gov/alsj/a11/a11.html)

[65] Datenblatt Biogon f-5.6/60mm (https://www.hq.nasa.gov/alsj/Biogon5.6_60mm_ZEISS.pdf)

[66] Preliminary Map of EVA Photographs and Television Pictures (https://www.hq.nasa.gov/alsj/a11/a11photomap.gif)

[67] Sun Angles (https://www.hq.nasa.gov/alsj/alsj-sunangles.html)

[68] Vasavada et al.: Lunar equatorial surface temperatures and regolith properties from the Diviner Lunar Radiometer Experiment; JOURNAL OF GEOPHYSICAL RESEARCH, VOL. 117, E00H18, doi:10.1029/2011JE003987, 2012

[69] Apollo Lunar TV Camera, Operations Manual (https://history.nasa.gov/alsj/a11/a11TVManual.pdf)

[70] https://www.history.nasa.gov/alsj/a11/as11-5864-69.jpg

[71] (Schweizer) Signalisationsverordnung (https://www.admin.ch/opc/de/classified-compilation/19790235/201701150000/741.21.pdf)

[72] Apollo 11 Video Library, One Small Step (https://www.hq.nasa.gov/alsj/a11/video11.html#Step) bei 109:42:28

[73] One Small Step (https://www.hq.nasa.gov/alsj/a11/a11.step.html)

[74] One Small Step (Version November 2017) (http://web.archive.org/web/20171116192006/http://www.hq.nasa.gov:80/alsj/a11/a11.step.html)

[75] Neue Zürcher Zeitung (NZZ) (http://www.hls-dhs-dss.ch/textes/d/D48585.php)

[76] LUNAR MODULE QUICK REFERENCE DATA (https://www.hq.nasa.gov/alsj/LM04_Lunar_Module_ppLV1-17.pdf)

[77] APOLLO LUNAR DESCENT AND ASCENT TRAJECTORIES (https://www.hq.nasa.gov/alsj/nasa58040.pdf)

[78] Apollo 11 Lunar Orbit Phase (https://history.nasa.gov/SP-4029/Apollo_11g_Lunar_Orbit_Phase.htm)

References

[79] Apollo 11 Mission Overview (https://www.hq.nasa.gov/alsj/a11/a11ov.html)

[80] LADEE (https://www.nasa.gov/mission_pages/ladee/main/index.html)

[81] NASA Completes LADEE Mission with Planned Impact on Moon's Surface (https://www.nasa.gov/ames/nasa-completes-ladee-mission-with-planned-impact-on-moons-surface)

[82] How Long Does It Take to Get to the ISS? (https://www.nasa.gov/audience/foreducators/topnav/materials/listbytype/TSD_Launching_Video.html)

[83] Raumfahrer erreichen ISS in nur sechs Stunden (http://www.spiegel.de/wissenschaft/weltall/sojus-rekord-raumfahrer-erreichen-iss-in-nur-sechs-stunden-a-891680.html#spCommentsBoxPager)

[84] Wikipedia: Kurs (Dockingsystem)

[85] The Apollo-Soyuz Mission (https://www.nasa.gov/mission_pages/apollo-soyuz/astp_mission.html)

[86] Soyuz rendezvous and docking explained (https://www.youtube.com/watch?v=M2_NeFbFcSw)

[87] Wikipedia: Gemini 11

[88] Wikipedia: Trident (SLBM)

[89] Apollo 17: Return to Orbit (https://www.hq.nasa.gov/alsj/a17/a17.launch.html): MPEG Clip just after 188:01:25

[90] ALSJ: The Return to Orbit (https://www.hq.nasa.gov/alsj/a11/a11.launch.html)

[91] Apollo 11 Recovery (https://www.youtube.com/watch?v=6fkl2tDO58s)

[92] APOLLO 11 The Fifth Mission: (https://history.nasa.gov/SP-4029/Apollo_11a_Summary.htm)

[93] Apollo 11 Astronauts Relax Following Successful Mission (https://www.nasa.gov/content/apollo-11-astronauts-relax-following-successful-mission)

[94] APOLLO LUNAR QUARANTINE PROGRAM (https://www.jsc.nasa.gov/history/oral_histories/McCollumGW_BogardD/ApolloLQP.pdf)

[95] 50 years ago, on the way to the Moon...
(https://www.nasa.gov/feature/50-years-ago-on-the-way-to-the-moon)

[96] Wikipedia (English): Apollo 11

[97] Apollo 11 Post Flight Press Conference
(https://www.hq.nasa.gov/alsj/a11/a11PostFlightPressConf.html)

[98] Wikipedia: Michael Collins (Astronaut)

[99] Wikipedia: Buzz Aldrin

[100] «Buzz Aldrin hat eine echte Begeisterung für die Schweiz»
(https://www.migrosmagazin.ch/archiv/aldin-hat-echte-begeisterung-fuer-schweiz)

[101] Buzz Aldrin Punches Guy - NEW - HD - READ BELOW (YouTube von Bart Sibrel)
(https://www.youtube.com/watch?v=OROlF8zB9z0)

[102] Mondlandung Hoax – Buzz Aldrin schlägt Reporter in das Gesicht
(https://wissenschaft3000.wordpress.com/2016/08/26/mondlandung-hoax-buzz-aldrin-schlaegt-reporter-in-das-gesicht/)

[103] Astronauts Gone Wild
(https://www.youtube.com/watch?annotation_id=annotation_3632523271&feature=iv&src_vid=OROlF8zB9z0&v=Qr6Vcvl0OeU)

[104] Wikipedia: Neil Armstrong

[105] Apollo 11 25th Anniversary - The White House (July 20th, 1994) (https://www.youtube.com/watch?v=Znyx2gTh3HU)

[106] Neil Armstrong - last interview - meets Alexej Leonov
(https://vimeo.com/80716491)

[107] Waren Sie wirklich auf dem Mond, Mister Armstrong?
(https://www.bild.de/news/2010/waren-sie-wirklich-auf-dem-mond-13525542.bild.html#fromWall)

[108] Warum schwieg Neil Armstrong
(http://www.bakonline.ch/2011/artikel254.php)

[109] Neil Armstrong will nicht schwören, dass er auf dem Mond war (https://www.youtube.com/watch?v=E4oaYqzE9Cc)

References

[110] Nachlass vom ersten Mann auf dem Mond bringt Millionenerlös (https://www.nzz.ch/panorama/neil-armstrong-sein-nachlass-bringt-millionen-ld.1433865)

[111] Apollo 13 (https://www.nasa.gov/mission_pages/apollo/missions/apollo13.html)

[112] Apollo 13 Re-entry (1970) (https://www.youtube.com/watch?v=wX8-Vmys-Fk) ab 42:00

[113] AS14-67-9361 (https://history.nasa.gov/alsj/a14/images14.html#Mag67) Die Bezeichnung des Handwagens ist MET (Modular Equipment Transporter)

[114] The Apollo 15 Hammer-Feather Drop (https://nssdc.gsfc.nasa.gov/planetary/lunar/apollo_15_feather_drop.html)

[115] Wikipedia: Raumstation

[116] Wikipedia: Liste der bemannten Raumflüge

[117] 40 Years Ago, Skylab Paved Way for International Space Station (https://www.nasa.gov/content/40-years-ago-skylab-paved-way-for-international-space-station)

[118] Launch of the Uncrewed Skylab Station (https://www.nasa.gov/mission_pages/skylab/missions/skylab1.html)

[119] Skylab 2: "We can fix anything!" (https://www.nasa.gov/feature/skylab-2-we-can-fix-anything)

[120] From Skylab to Station: Out of This World Science (https://www.youtube.com/watch?v=nmWbm9ab4n4)

[121] Skylab Videos - Skylab: The First 40 Days (https://www.nasa.gov/mission_pages/skylab/videos)

[122] Skylab Videos - The Skylab Legacy -- Long Duration Space Flight (https://www.nasa.gov/mission_pages/skylab/videos)

[123] BBC: Abenteuer im Weltraum - Die grossen Missionen der NASA; 4 DVD

[124] SP-401 Skylab, Classroom in Space (https://history.nasa.gov/SP-401/ch3.htm)

[125] Wikipedia: Skylab

[126] Wikipedia: Apollo-Sojus-Test-Projekt
[127] Wikipedia: Space Shuttle
[128] Space Shuttle (https://www.nasa.gov/returntoflight/system/system_STS.html)
[129] This Week in NASA History: STS-71 Docks with Mir – June 29, 1995 (https://www.nasa.gov/centers/marshall/history/this-week-in-nasa-history-sts-71-docks-with-mir-june-29-1995.html)
[130] Wikipedia: STS-91
[131] Wikipedia: STS-51-L
[132] Wikipedia: STS-107
[133] Space Shuttle Discovery Landing (STS-119) (https://www.youtube.com/watch?v=YIAwP3EGong)
[134] Wikipedia: STS-119
[135] Wikipedia: Sandra Magnus
[136] Using the Space Gym With Samantha Cristoforetti | ISS Video (https://www.youtube.com/watch?v=9P0AN1N_xyQ)
[137] Wikipedia: Spacelab
[138] Space Shuttle – Spacelab (https://www.esa.int/Our_Activities/Human_Spaceflight/Space_Shuttle/Spacelab)
[139] Wikipedia: Mir (Raumstation)
[140] Wikipedia: Wiedereintritt
[141] NASA - Benefits of the Shuttle-Mir Program (https://www.nasa.gov/audience/formedia/factsheet/shuttle_mir_factsheet.html)
[142] Wikipedia: Sarja
[143] Space Shuttle (https://www.nasa.gov/mission_pages/shuttle/shuttlemissions/sts119/multimedia/fd11/fd11_gallery.html)
[144] Wikipedia: Internationale Raumstation
[145] Heavens Above (https://www.heavens-above.com/main.aspx)
[146] Wikipedia: Shenzhou 5
[147] Wikipedia: Tiangong 1

References

[148] Wikipedia: Tiangong 2

[149] Apollo 11 Preliminary Science Report (1969) (https://history.nasa.gov/alsj/a11/a11psr.html)

[150] «Touristen werden auf den Mond fliegen» (https://www.blick.ch/life/zurbuchen-ueber-kommerzielle-weltallfluege-touristen-werden-auf-den-mond-fliegen-id8768155.html)

[151] Fake Dutch 'moon rock' revealed (http://news.bbc.co.uk/2/hi/8226075.stm)

[152] «Weltraumforschung ist meistens Teamwork» (https://www.derbund.ch/bern/dossier/die-samstagsinterviews/Weltraumforschung-ist-meistens-Teamwork/story/30158617)

[153] Weltraumforschung – Medizintechnik (http://www.unibe.ch/unibe/portal/content/e796/e800/e10902/e310398/e362685/up_140_heft_ger.pdf)

[154] Lunar Reconnaissance Orbiter (https://www.nasa.gov/mission_pages/LRO/main/index.html)

[155] (https://www.nasa.gov/images/content/369233main_lroc_apollo11_256x256.jpg)

[156] LRO Gets Additional View of Apollo 11 Landing Site (https://www.nasa.gov/mission_pages/LRO/multimedia/lroimages/lroc_20091109_apollo11.html)

[157] Lunar Laser Ranging: A Continuing Legacy of the Apollo Program (https://www.hq.nasa.gov/alsj/LRRR-94-0193.pdf)

[158] A Critical Review of the Lunar Laser Ranging (https://arxiv.org/abs/1805.05863)

[159] Unglaubliche Ausdauer: Mars-Rover ist seit 15 Jahren in Betrieb (https://www.bluewin.ch/de/news/wissen-technik/unglaubliche-ausdauer-mars-rover-ist-seit-15-jahren-in-betrieb-119251.html)

[160] LADEE PRESS KIT (https://www.nasa.gov/sites/default/files/files/LADEE-Press-Kit-08292013.pdf)

[161] Wikipedia: Chronologie der Mondmissionen

[162] Vertrag über die Grundsätze zur Regelung der Tätigkeiten von Staaten bei der Erforschung und Nutzung des Weltraums einschliesslich des Mondes und anderer Himmelskörper (https://www.admin.ch/opc/de/classified-compilation/19670016/index.html)

[163] Back to the Moon the sustainable way (http://exploration.esa.int/moon/60261-back-to-the-moon-the-sustainable-way/)

[164] Wikipedia (English): List of missions to the Moon

[165] Neil Armstrong gave her a vial of moon dust, she says. She's suing so NASA won't take it. (https://www.washingtonpost.com/news/speaking-of-science/wp/2018/06/12/neil-armstrong-gave-her-a-vial-of-moon-dust-she-says-shes-suing-so-nasa-wont-take-it/?utm_term=.b140435a3f4f)

[166] David Icke: DER LÖWE ERWACHT, Zweite Auflage, 2011, ISBN 978-3-928963-45-9

[167] Leslie Kean: UFOs; ISBN 978-3-86445-025-9

[168] Wikipedia: Sacsayhuamán

[169] Wikipedia: Liste der größten Monolithen der Welt

[170] Cathy O-Brien und Mark Phillips: Die TranceFormation Amerikas, ISBN 978-3-928963-05-3 (https://www.orellfuessli.ch/shop/home/artikeldetails/ID6240000.html?ProvID=10917735)

[171] Budget of NASA (https://en.m.wikipedia.org/wiki/Budget_of_NASA)

[172] Air Force is spending more on space, but modernization path still a big question (https://spacenews.com/air-force-is-spending-more-on-space-but-modernization-path-still-a-big-question/)

[173] United States Space Force (https://en.m.wikipedia.org/wiki/United_States_Space_Force)

[174] JFK Secret Societies Speech (full version) (https://www.youtube.com/watch?v=zdMbmdFOvTs)

References

[175] John F. Kennedy, Rede im Waldorf Astoria Hotel, New York City, vom 27. April 1961
(http://www.grundrechtsschutzinitiative.de/95e5cd090ea0d0fabb01134df77f3aa6_J%20F%20Kennedy%20Rede%2027%20Apr%201961.pdf)

[176] Kubricks geheime Botschaften
(https://www.dietiefe.com/2016/03/15/die-unheimlichen-botschaften-des-meisterregisseurs/)

[177] Wikipedia: Shining (1980)

[178] Wikipedia: Orson Welles

[179] ZEIT ONLINE: Original-Aufnahmen bleiben wohl für immer verschollen
(https://www.zeit.de/online/2009/30/mondlandung-aufnahmen-verschwunden)

[180] FOCUS ONLINE: Die verlorenen Videoaufnahmen
(https://www.focus.de/wissen/weltraum/raumfahrt/mondlandung-die-verlorenen-videoaufnahmen_aid_417352.html)

[181] Model Helps Search for Moon Dust Fountains
(https://www.nasa.gov/mission_pages/LADEE/main/lhg.html)

[182] Wikipedia: Verschwörungstheorien zur Mondlandung, Abschnitt *Der seitliche Schatten des Fotografen*, Version 15. Dezember 2013, 11:32
(https://de.wikipedia.org/w/index.php?title=Verschw%C3%B6rungstheorien_zur_Mondlandung&oldid=125455713#Der_seitliche_Schatten_des_Fotografen)

[183] Wikipedia: Keine Theoriefindung
(https://de.wikipedia.org/wiki/Wikipedia:Keine_Theoriefindung)

[184] Der seitliche Schatten des Fotografen – willkürlich gelöscht
(https://de.wikipedia.org/wiki/Diskussion:Verschw%C3%B6rungstheorien_zur_Mondlandung/Archiv/2014#Der_seitliche_Schatten_des_Fotografen_.E2.80.93_willk.C3.BCrlich_gel.C3.B6scht)

[185] Weblinks zu privaten Werbezwecken?
(https://de.wikipedia.org/wiki/Diskussion:Verschw%C3%B6rungstheorien_zur_Mondlandung/Archiv/2013#Weblinks_zu_privaten_Werbezwecken.3F)

[186] Shuttle Atlantis STS-132 - Amazing Shuttle Launch Experience (https://www.youtube.com/watch?v=5KygwcZ545U)

[187] Wikipedia: Mondlandefähre

[188] Wikipedia: Des Kaisers neue Kleider

[189] Wikipedia: Verschwörungstheorie

[190] Das Drama der Experten (https://www.tagesanzeiger.ch/ausland/die-tsunami-katastrophe/Das-Drama-der-Experten/story/21080894?dossier_id=885)

[191] Wikipedia: Bielefeld-Verschwörung

[192] 30-Second Reel of Building 7 Collapse Footage (https://www.youtube.com/watch?v=Mamvq7LWqRU)

[193] http://www.911thology.com/

[194] Warum musste John F. Kennedy sterben? (https://www.handelsblatt.com/politik/international/wilde-theorien-warum-musste-john-f-kennedy-sterben/9108670.html)

[195] Wikipedia: Roswell-Zwischenfall

[196] INTERNATIONAL UFO MUSEUM AND RESEARCH CENTER (https://www.roswellufomuseum.com/)

[197] Wikipedia: High Frequency Active Auroral Research Program

[198] Die Wahrheit über die Kondensstreifen (https://www.tagesanzeiger.ch/wissen/technik/die-wahrheit-ueber-die-kondensstreifen/story/27639462)

[199] Kennen Sie eigentlich die ENMOD-Konvention gegen Wetterwaffen aus dem Jahre 1976? (https://www.radio-utopie.de/2011/03/11/kennen-sie-eigentlich-die-enmod-konvention-gegen-wetterwaffen-aus-dem-jahre-1976/)

[200] «Glyphosat ist krebserregend» (https://www.beobachter.ch/umwelt/flora-fauna/unkrautvertilger-glyphosat-ist-krebserregend)

[201] Netzwerk Impfentscheid (https://impfentscheid.ch/)

[202] Wikipedia: Ernst Cincera

References

[203] Merkel war es wirklich nicht (https://www.zeit.de/politik/ausland/2016-10/fluechtlingspolitik-fluechtlinge-angela-merkel-balkanroute-offene-grenze)

[204] Geldmengen (https://www.snb.ch/de/iabout/monpol/monstat/id/monpol_monstat_geldmengen)

[205] Mathias Binswanger: Geld aus dem Nichts, ISBN 978-3-527-50817-4

[206] Federal Reserve Bank of San Francisco (https://www.frbsf.org/education/publications/doctor-econ/2003/september/private-public-corporation/)

[207] B. Ray Hawke et. al.: The origin of lunar crater rays (Icarus 170 (2004) 1–16)

[208] Wikipedia: Libration

[209] Wikipedia (English): Far side of the Moon

[210] Rupert Sheldrake: Der Wissenschaftswahn, ISBN 978-3-426-29210-5

[211] Apollo program (https://wikispooks.com/w/index.php?title=Apollo_program&redirect=no)

[212] Wikipedia: Mond

[213] Wikipedia: Mondbahn

[214] Stellarium 0.15.2, (http://stellarium.org/)

[215] Wikipedia Aggregat 4 (https://de.wikipedia.org/wiki/Aggregat_4)

[216] Wikipedia Schaltsekunde (https://de.wikipedia.org/wiki/Schaltsekunde)

[217] NASA Ranger (https://science.nasa.gov/missions/ranger/)

[218] NASA Apollo 11 (https://www.nasa.gov/mission_pages/apollo/apollo11.html)

[219] Apollo 11 Crew Meets With President Obama (https://www.nasa.gov/multimedia/imagegallery/image_feature_1422.html)

[220] Apollo 40th Anniversary Press Conference (https://www.nasa.gov/image-feature/apollo-40th-anniversary-press-conference)
[221] Star Facts (https://walkoffame.com/special-apollo-xi/)
[222] Blog in the web archive: (http://web.archive.org/web/20200729102819/https:/fellowshipoftheminds.com/are-the-crew-members-of-1986-space-shuttle-challenger-still-alive)
[223] Michael J. Smith (https://en.wikipedia.org/wiki/Michael_J._Smith)
[224] Profile of Michael J. Smith (https://directory.engr.wisc.edu/ie/faculty/smith_michael)
[225] ISS Astronaut Talks About Life in Space with BBC, NASA, 2:10ff (https://www.youtube.com/watch?v=C8ZWjdP74aA)
[226] Die Raumfahrt hat ein Schimmelproblem (https://www.wissenschaft.de/astronomie-physik/die-raumfahrt-hat-ein-schimmelproblem/)
[227] Erfahrungen mit Russlands MIR-Station (https://www.esa.int/Space_in_Member_States/Germany/Schutz_gegen_blinde_Weltraum-Passagiere)
[228] mRNA Vaccine Toxicity (https://doctors4covidethics.org/mRNA-vaccine-toxicity/)
[229] The deep hot biosphere: the myth of fossil fuels / Thomas Gold; ISBN 978-0-387-95253-6

www.ingramcontent.com/pod-product-compliance
Lightning Source LLC
Chambersburg PA
CBHW052158220526
45471CB00004B/1718